# PRAISE FOR
# *TREES OF POWER*

"In these times of discouraging news, it is heartening when a book like *Trees of Power* comes to the fore. Informative, inspirational, and practical, this timeless work will draw the reader in wonder into the selflessly giving world of ten significant tree allies. Inside you'll find not only information *about* the trees, but instructions on how to collect and store seed and propagate trees through techniques such as grafting and cuttings. All for the purpose (to paraphrase the author) of inspiring us to follow our inspirations, not our fears. Not to *hope* for change, but to *create* change!"

—MARK SHEPARD, author of *Restoration Agriculture*

"*Trees of Power* turned out to be so much more than I expected. The chapters on growing and propagating trees will be of great use to both amateurs and professionals wanting to grow woody plants, written as they are by a commercial nursery owner who really knows his stuff. The ten chapters that cover some of the most useful trees to grow glow with Silver's enthusiasm and knowledge."

—MARTIN CRAWFORD, founder, Agroforestry Research Trust

"The underlying message of this book is crystal clear: Anyone can grow their own trees. Akiva Silver's writing is refreshingly straightforward and full of encouragement, offering readers a humble vision of connecting our lives more closely with trees. Emphasizing keen observation, a love for the local landscape, and a toolkit of simple propagation and planting skills, Akiva gives readers little to do after reading this book but get on with planting!"

—STEVE GABRIEL, author of *Silvopasture*

"*Trees of Power* is an intensely practical book that doesn't read like a textbook. It is a passionate testament to Akiva Silver's arboreal friends and a joyous celebration of how he works with nature's terrestrial giants to earn a living. Akiva assumes little existing knowledge. He offers tips to improve your tree identification skills and covers the full range of tree cultivation techniques. There is plenty in here for the experienced orchardist. However, even if you have no intention of rolling up your sleeves and joining him in the orchard but simply love trees, this book is for you, too."

—BEN RASKIN, Head of Horticulture, Soil Association

"From shoots to nuts, Akiva Silver provides helpful details on every aspect of each of the trees in the book. The lyrical writing makes me, after forty years in California, miss the deciduous forests of Missouri. I hope there will be a sequel, one that covers even more than ten arboreal allies."

—ROBERT KOURIK, author of
*Designing and Maintaining Your Edible Landscape Naturally*

"Certainly *treeculture* could have been a word in its own right. Trees are sentient beings charged with regulating the breath of this living planet. Contemplating the ten important species spotlighted in *Trees of Power* has me excited to keep on planting. Here are trees worthy of every regenerative landscape. Trees that provide us with food, nutrient cycling, fungal havens, and diverse ecosystems that support yet even more life. In the classic *Tree Crops*, J. Russell Smith spoke about how farming should fit the land, how crop-yielding trees can fill in all those nooks on challenging terrain. Akiva Silver's insights into the essentials of arboreal care provide the how-to guide to successfully establish those tree friends. Take to heart his tree wisdom, and make your special place complete."

—MICHAEL PHILLIPS, author of
*The Holistic Orchard* and *Mycorrhizal Planet*

"More than just a how-to book, *Trees of Power* offers a glimpse into an inspired way of working from someone who has found his calling. Whether you want inspiration or information, this handbook is a simple and profound gem about being a force for positive change. Akiva uses trees as his medium, but the lessons here are timeless and translate into any other context."

—BEN FALK, author of
*The Resilient Farm and Homestead*

# Trees of Power
## Ten Essential Arboreal Allies

Akiva Silver
*foreword by* **Samuel Thayer**

Chelsea Green Publishing
White River Junction, Vermont
London, UK

Copyright © 2019 by Akiva Silver.
All rights reserved.

Unless otherwise noted, all photographs copyright © 2019 by Akiva Silver.

No part of this book may be transmitted or reproduced in any form
by any means without permission in writing from the publisher.

Project Manager: Alexander Bullett
Developmental Editor: Benjamin Watson
Copy Editor: Laura Jorstad
Proofreader: Katherine Kiger
Indexer: Shana Milkie
Designer: Melissa Jacobson

Printed in the United States of America.
First printing February, 2019.
10 9 8 7 6 5 4 3     23 24 25

**Library of Congress Cataloging-in-Publication Data**
Names: Silver, Akiva, author.
Title: Trees of power : ten essential arboreal allies / [Akiva Silver].
Description: White River Junction, Vermont : Chelsea Green Publishing, [2019]
   | Includes bibliographical references and index.
Identifiers: LCCN 2018049017| ISBN 9781603588416 (pbk.) | ISBN 9781603588423 (ebook)
Subjects: LCSH: Trees. | Trees--Propagation. | Tree planting.
Classification: LCC SD391 .S52 2019 | DDC 634.9--dc23
LC record available at https://lccn.loc.gov/2018049017

Chelsea Green Publishing
85 North Main Street, Suite 120
White River Junction, VT 05001
(802) 295-6300
www.chelseagreen.com

To my children,
may you love the Universe as much as it loves you.

# CONTENTS

Foreword — ix
Preface — xi

**PART ONE**

## Concepts and Skills — 1

1. Life Fountains — 3
2. Planting — 33
3. Propagation from Cuttings — 67
4. Propagation from Seed — 77
5. Propagation by Grafting — 86
6. Propagation by Layering — 95

**PART TWO**

## The Allies — 101

7. Chestnut: The Bread Tree — 103
8. Apple: The Magnetic Center — 135
9. Poplar: The Homemaker — 157
10. Ash: Maker of Wood — 165
11. Mulberry: The Giving Tree — 173
12. Elderberry: The Caretaker — 185
13. Hickory: Pillars of Life — 194
14. Hazelnut: The Provider — 209
15. Black Locust: The Restoration Tree — 223
16. Beech: The Root Runner — 235

Afterword: Leaves of the Same Tree — 245

Acknowledgments — 249
Exercises for Increasing Awareness — 251
Recommended Reading — 255
Resources for Plant Material — 257
Notes — 259
Index — 261

# FOREWORD

There is a cemetery where I like to gather acorns, high on a sandy hill rising near the cool blue waters of Lake Superior. The grass is so scant on this poor soil that it scarcely needs mowing. There are no houses in sight; surrounded by forest on three sides and an abandoned apple orchard on another, this place is quiet, even for a cemetery. In October the acorns begin to fall from the spreading limbs of the scattered oaks, covering the ground in a layer of gems that almost obscures the mosses and hawkweeds. Bears and deer come nightly to indulge in the bounty, squirrels and jays by day—but most years there are more than enough acorns for my pantry, too.

As I fill my basket, I stop to read the headstones and pay respect to bygone strangers whose familiar surnames live on through local businesses, descendants, and road names. Most were Scandinavian immigrants who came here to clear trees, hoping to eke out a living from field crops on the sterile sandy uplands or the red clay lowlands. The soil had other plans.

A cemetery is a good place to wonder. Why do people who toil to destroy trees in life still want to be comforted by them in death? Trees speak to our souls because they offer life to our bodies, a timeless proposition that predates and outlasts us. Trees connect us to forever. We seem to understand that on the deepest of levels, but our timing is off. We need to understand it on the shallow level, too, and redeem the offer while it's still valid. That's what this book is about.

I met Akiva Silver through the kind of lucky hunch that is often called destiny in retrospect. His Twisted Tree Farm was holding its annual Nut Bonanza—an educational symposium on the gifts and economic possibilities of native nut trees—the same week that I was going to be in New York State. I had to be there; a few emails later and I was enlisted to help. I pulled into the gravel driveway just before midnight, to the comforting rush of creek water, my beat-up Corolla weighted down with bushels of supplies. On the other side of the porch light was an energetic man beaming with determination and insight, whose impassioned speech was inflected by the love of trees. We stayed up later than we should have, trading hopes and tales of nut trees, fruit trees, shade trees, medicine trees, firewood trees, climbing trees, before finally sinking to sleep on a floor made from trees, under a roof made from trees, in a house heated with trees.

Trees are like that: almost everywhere, in almost everything, mostly overlooked. They give so generously that we take their gifts for granted. But Akiva does not—he has made it his lifework to proclaim the contract of trees: *I will make it if you pick it; I will drop it if you pick it up.*

Akiva's book is about signing that contract—committing the deed of planting trees—and then carrying out our part of the agreement. It is an optimist's manual of solutions—not easy or quick duct-tape fixes to environmental problems, but real, good, long-term answers to the question of how to be fully and vibrantly human in a world of Nature. It is not a map to the high end of a sinking *Titanic*, nor a cry to bring out the lifeboats. We are not aboard a doomed ship, bobbing in a frigid sea; we are on Earth, exactly where we are supposed to be, surrounded by all that we need, in a garden that wants tending.

Akiva and I know that work well: the dirty fingers and worn hands of a planter, transplanter, pruner, thinner, mulcher, weeder. This is followed by the heart-pounding thrill of harvest—filling sacks of nuts, baskets of apples, tubs of berries, barrels of sap, boatloads of wild rice. Next comes the deep satisfaction of turning that harvest into food: bottles of delicious nut oils, golden maple syrup, cider and applesauce, fruit leather, flour. This wholesome tending to the process of life leaves a warm glow in our hearts, the deep comfort of living in a world of life-giving trees. Yes, there is work to do—marvelous work, the kind that to chipmunks and orioles is just life.

Trees beckon us to sit at their feet, humbly, and listen. They speak of the supposedly distant past, reminding us that it was scarcely more than yesterday. They link us to a future that becomes, through them, imaginable, almost palpable. Perhaps we cannot guess what the future holds, but we can *plant* it. We do not have to be shortsighted just because we are short. Trees are the answer to so many of our ills, and the ladder to so many of our dreams. They are the arms and hands of the Earth, reaching up to the heavens on our behalf, grasping the slippery currency of sunlight and rendering it, through their wondrous alchemy, into the stuff of life— our life and theirs. All the trees ask in return for that gift is that we live and work among them. Their leaves whisper of an alternative economy, serving different values, that will be here, scarcely more than tomorrow.

—*Samuel Thayer*

# PREFACE

My friend Mark and I paddled down the Clarion River in Pennsylvania. We were dressed in full buckskin. Our clothes were made from hides we had brain-tanned ourselves. We carried longbows of hickory and ash that we had made. Flint-tipped arrows of viburnum wood filled our quivers. Our minds were filled with vision. To gather all of our food, live in a shelter we made without tools, sit by a fire lit by no match, and become one with the wilderness. We despised civilization and revered nature. Every morning and evening was a fully attuned meditation to the forest around us. We had trained for this trip for years, learning ancient primitive skills, spending thousands of hours in the woods.

Our camp was far from where any hiker would discover us, tucked back on the mountain under a canopy of rhododendron and red maple. It was the month of May. I had just left Rochester, New York. Living in the suburbs, I was craving the wilderness, desperate for her truths. The town was dirtied everywhere by the hands of people. Houses, wires, fences, garbage, streets, electric lights, cars: It was all in the way of what I thought was real.

As the days went by on our camping trip, I slowly began to realize how quiet it was there. It was too quiet. When I left Rochester, it had been bursting with the life of spring. The dawn chorus of birds had been overwhelming during my morning sits. But here in the wilderness, under an endless canopy of red maple, it was silent. Maybe I would see a robin or two at dawn, maybe a chipmunk. In Rochester, in the heart of the suburbs, I had been encountering thousands of birds, foxes, raccoons, deer, mink, opossums, skunks, squirrels, coyotes, and many other creatures on a daily basis. Here in the "wilderness," it was silent.

This was the beginning of my realization that people are not bad. We can be helpful or destructive to wildlife populations. It all depends on how we focus our energy, on what we do to the soils and how we influence plant communities.

The hills along the Clarion River where we camped were covered in close to 100 percent red maple. Those red maples had seeded in at just the right time following a heavy logging operation 50 to 80 years ago. If someone had taken the time to plant just a few specific trees at the time of disturbance, then I would have been in a very different forest. Leaving that land alone following disturbance had its own dramatic effect. Choosing to do nothing with a piece of land is a big choice that carries significant consequences.

We live at a time where there is widespread disturbance all around us. The ground is open and waiting for seeds. We can bemoan the tragedies that nature has endured or we can cast seeds and plant a future. We can and do influence the ecosystems around us more than any other species. That influence can come through reckless destruction, blind abandonment, or conscious intent. This book is about making the choice to participate in nature through conscious intent by working with trees.

Because we have chain saws and bulldozers at our disposal, and because trees cannot run, hide, or fight back, they can be thought of as weak, defenseless organisms. The truth is that trees are resilient beings that have been around for hundreds of millions of years, enduring shifts in climate, being chewed on and trampled by everything from voles to elephants. They have come from the mysteries of the deep past. When I look at the sky, what I see there is not simply blue. There's a radiance, an energy, a power. It is from this power that trees feed. Literally building their bodies out of the radiant sky, trees of power are strong beings to ally ourselves with. Their wisdom and abilities are very different from our own. They are mysterious organisms that naturally fit into a symbiosis with us if we can learn to work with them.

# PART ONE

# Concepts and Skills

When I was a kid growing up in the suburbs, my life was a shuttle among school, stores, sports, and TV. I had no idea that nature was breathing aliveness all around me. I certainly had no idea of how to work with nature or what I would even do. This first part of this book is about cultivating a relationship with the natural world both conceptually and practically.

We all have a lot to learn about living on this Earth. It is a strange and wild place with endless nuance and variation. As soon as we learn something, we find more questions. Life is dynamic, requiring awareness and adaptability. The skills and concepts presented here are just a base. You can use them to further develop your own ideas and skills as you respond to the present moment and the changing world around you.

CHAPTER ONE

# Life Fountains

Trees in fog stand without leaves, dark stems in a maze of inexhaustible intricacy. Patterns laid upon patterns in a seeming randomness that gives way to a single beautiful scene. These life fountains spring from the ground, rising from a dark and mysterious world fully charged with life. They rise and rise and then spread. From the end of every branch tip drip the fountains. Seeds rain down, feeding birds and mammals. We breathe these trees through our lungs, shelter ourselves with their wood, and fill our bodies with the energy of their fruit.

These fountains of life are incredible beings that perform so many services for free and indefinitely. They have the ability to reproduce themselves, run on sun and rain, build wood out of carbon in the sky, create flavors, carbohydrates, proteins, fats, medicine, and vitamins. We are just tiny animals scampering beneath them, picking up their gifts as fast as we can, because there is not enough time to keep up with the rain of presents. The feel of autumn in the wind pushes us to gather faster, filling bucket after bucket. The harvest looks staggering. It fills trucks and porches. *Where will we put it all?* and *How will we have time to process all this?* are some of the thoughts we have, and still there is so much more lying on the ground. Millions of pounds in my county alone.

*"Mind if I gather nuts off your lawn?"* They are waste to my culture; it's a chore to rake them up off the grass. The gifts of the trees, of the Universe, are largely ignored. It is a strange world indeed. I can't explain the physical joy I feel filling buckets with nuts. Crawling around on my hands and knees surrounded by a staggering abundance, I sometimes laugh out loud like a madman and look around to see if anyone heard me. Sometimes a guy on a bicycle stares at me. But I have no time to worry about that. It is the harvest season and I am flying high. I need to keep reminding myself to stay calm.

# Trees of Power

The heart of the gatherer is one of gratitude and amazement. I have been astounded so many times harvesting. As I start to pick up the first bushels of wild pears, I realize just how much is there. I sell wild pears to a cidery that presses them into perry (pear wine, which is a very excellent drink with a long history in Europe). Last fall my family and a friend gathered over 3,000 pounds of wild pears from a handful of trees in two days. Over 1,000 pounds came from a single tree. Seeing that much fruit does something to you. It is impossible not to be impressed even if you aren't interested in pears. But we are interested; it's a part of our livelihood, and each bushel is cash. We gather with speed and efficiency, sometimes chatting, sometimes working silently. It is good work, work our bodies and minds were built for. At night we see pears when we close our eyes. We have a connection to those trees. We care what happens to them. To us, it seems like a good idea to plant more of them. The highest level of appreciation comes through participation.

I hope that this book inspires you to gather and plant. The trees here are some of the most enjoyable beings on Earth to work with. If you watch for them, they will overwhelm you sometimes. It will seem like they are merely offering you thousands of pounds of food and seed for free, but they have their own interests at heart. By taking from them, you will be helping them. You will be partnered. Your work on this world does not have to be drudgery or bad for the planet. By working with trees we can find abundance and spread it.

Life circles around trees; it is drawn in like a magnet. One crab apple tree in the middle of winter will pull in birds, possums, mice, deer, raccoons, wild children, and countless other forms of life. Animals and people will travel for miles to gather persimmons and chestnuts. Songbirds will flock to mulberries. These are magnetic trees, fountains of life that shower the Earth with abundant gifts. When we become aware of these trees, we can begin to work with them and elevate the level of abundance in our world to staggering heights.

Humans can have a positive influence on nature. We can enhance ecosystems to the benefit of ourselves and wildlife at the same time. I see a world filled with endless opportunity. There are gifts falling down all around us. Many folks don't see them at all, even while they are taking the time to pick up these presents and throw them away. This book is a guide and a catalyst. I hope that it helps you realize there is good work to do everywhere and that you can be a positive force for nature and for yourself. You can harvest food and medicine, make money, breathe gratitude, and leave beauty in your wake by working with trees. They are filled with power, and that power is freely offered to us. Partnering with trees is as natural as breathing. We inhale their exhalations and they inhale ours. We are designed to work with each other.

The trees in this book are my allies. They feed me, keep me warm, provide money, shelter, medicine, and tools. These beings that feed on light

This clover plant can deposit nitrogen and carbon deep into the soil a lot easier than I can.

do amazing work. Fruit, nuts, flowers, shade, wildlife, and wood—trees are offering, always offering. Stretched toward the sky, rooted into the earth, they offer a partnership. I think you will find, if you work with trees, that they are extremely generous beings. You will find yourself showered in more abundance than you are able to receive. I would rather partner with trees than any bank, institution, or lawmaker. These trees are my allies and they can be yours, too.

Every seed, cutting, or small tree that you ever hold in your hands wants to live. It wants the same thing you do. You are its ally as much as it yours. You are able to see and do things that are not possible for the plant. Humans can be amazing helpers to the plants we choose to work with. Alliances work both ways.

Where there is a dry rocky soil, we can change that. Where there are strip mines, eroded hillsides, or poisoned ground, we can add plants, who can do the healing work that's needed. They will do the work of bringing things back, they will build the soil and feed the birds. They can make it all happen, and they will, with or without us. The plants are the stewards of the Earth, taking care of all the animals, feeding us all.

If you want to add carbon to the soil, you can dig a hole and shovel it in. You can also let roots extend down through cracks in the rock and deposit carbon for you. Let the plants do the work. They want to.

## Tree Crops and Soil Carbon

In 1929 J. Russell Smith wrote a very powerful book called *Tree Crops*. It is about how trees can provide the crops that annuals do. Crops like corn, wheat, and soybeans can be replaced or supplemented with oak, chestnut, hazel, and hickory. Life can continue from year to year when the crops are grown on trees. The soil is saved and built upon. In our current system of annual agriculture, life is erased at least once a year. Just look at a field after it's been plowed—it could not be made more barren. We can do better, and that is what *Tree Crops* is all about.

In Smith's book the trees are given grain equivalents. For example, chestnuts are a corn tree, kiawe trees are stock feed trees, walnuts are meat and butter trees, and so on. This concept has inspired my work and is a huge reason why I am writing this book. If J. Russell Smith were alive today, I would love to thank him. His book and concepts are revolutionary. Grain growing on trees sounds simple, but it is remarkably powerful in practice.

The benefits that tree crops can provide are multifaceted. They are far-reaching, powerful benefits that have an effect on wildlife, soil, water, nutrients, climate, and even our human consciousness.

When we dig up a patch of ground with a shovel or nine-bottom plow, a lot happens. Air is brought into the soil rapidly. Microbial activity skyrockets. Carbon is burned off very fast—it literally vaporizes into the atmosphere. Erosion and climate change are the biggest problems created by annual agriculture. J. Russell Smith offered a solution back in 1929. His book not only speaks of philosophy but also gives examples of people around the world using tree crops for thousands of years.

Changing the color of a field from green to brown has enormous consequences. There is a reason that nature covers the soil. When rain strikes bare soil, it compacts it, changing its structure to create a more impervious layer. This leads to runoff and erosion. When bare soil is exposed to air, the organic matter becomes volatilized. Organic matter is essentially carbon, and when carbon meets oxygen, carbon dioxide is formed. $CO_2$ is light enough to float away. Every time a field is plowed, $CO_2$ floats up into the atmosphere. This is detrimental not only to the climate, but also to soil.

Scientists now debate which leads to more greenhouse gases in the atmosphere, transportation or tillage. Maybe you don't believe what scientists say. I often don't: They seem to change their paradigm every couple of generations. What I do know is that carbon is burned off when its exposed. I can see this in a rotting compost pile that shrinks and shrinks over time. It's not sinking into the soil and breaking down; it's actually vaporizing into the air. Some nutrients leach out into the soil, but many more actually vaporize. You can even smell the loss of nitrogen and other nutrients around some compost piles and around most farms. It is fully understood and accepted that plowing fields is a huge factor in climate change as, by its very nature, tillage burns carbon. When we drive by plowed-up acres of brown dirt, we are looking at a piece of land that's far removed from nature's design. The soil structure is being destroyed and massive amounts of carbon are heading up into the sky.

There are alternatives, beautiful ones that make economic and ecological sense. The soil's capacity to store and build carbon is highly underutilized by our current agricultural system, dependent as it is upon annual crops. I think of the soil as an enormous sponge that can soak up huge amounts of carbon.

All this extra carbon in the atmosphere comes from ancient forests that were buried and turned to coal and oil over time. It is the bodies of ancient trees we are burning in engines these days. This same carbon has been racing around and through the planet since forever. It is the plants who pull it out of the sky.

Carbon in the soil is translated as organic matter. Organic matter holds four times its weight in water, while at the same time it drains very well through a network of capillaries. Soils high in organic matter can absorb tremendous amounts of rain in a very short time period. They can supply moisture to plants evenly over long time spans. However, organic matter is vulnerable to exposure; it is easily washed away or volatilized into the atmosphere. New England soils from the late 1700s contained about 20 percent organic matter. These soils were totally resilient in their ability to deal with droughts and floods. As forests were cleared and land was plowed, organic matter was lost into the rivers and the atmosphere. Today it is common for agricultural fields to contain around 1 percent organic matter.

Carbon belongs in the soil. The cultivation of annuals leads to carbon moving into the atmosphere. The soil is where we can store carbon, where we can use it. The best way to put carbon back into the soil is through the roots of plants. Through their leaves, they suck carbon dioxide right out of the air. They use this carbon to build their bodies, above- and belowground. As plants and trees die, the carbon is released back into the air and soil.

We have heard some interesting solutions to climate change, everything from improved lightbulbs to launching thousands of mirrors into space to reflect the sun's rays. It's a complex world and I do not claim to have all the answers. What I do know is that the soil is an enormous sponge that can soak up all the carbon we burn. Adding carbon to the soil benefits plants, growers, and life in general. Tree crops are one method for increasing soil carbon—simple, cheap, and effective. Sometimes it's hard to see that the best solutions have been right in front of us the whole time.

If we want to slow down climate change, then we should burn less fossil fuels, but if we want to reverse it, then here is a way. Plants drink carbon. They put it down into their roots, deeper than you will ever reach with a shovel or a plow. Plants bury carbon. Bare soil burns it. Our system of annual agriculture is destroying plant and animal communities and altering the climate. I believe that in transitioning from annual agriculture to perennial plantings, we would not only slow climate change, but reverse it. I am not alone in this belief.

A 0.4 percent increase in soil carbon stocks each year would offset all the carbon we burn.[1] It really is that simple. We can store all the carbon we need to in the soil. It would be profitable to farmers, decrease drought and flood damage, and reverse climate change. Maybe you are asking, "Why don't we do that?" Some of us already are. We are not waiting for governments or universities. Many farmers are already working on this. They have taken the soil-carbon challenge and are seeing just how fertile they can make their lands. Reversing climate change is a by-product of sound agricultural practices.

# Life Fountains

Fields like this look normal to our trained eyes, but they are patches of land where ecology has been reduced to near zero, and where carbon and nutrients are being vaporized and leached out rapidly.

Tree crops offer an alternative agriculture of abundance, one that has been flourishing for the last few millennia. Here are planted chestnut trees in front of a stand of black walnut with the hillside covered in wild oak and hickory. Hemlock Grove Farm, West Danby, New York.

The carbon cycle is one of the most powerful forces on Earth. Understanding it is the key to climate change.

The loss of carbon in the soil happens simply from exposure to air. Rain and intense sun make it worse, but it happens very fast anywhere that the mulch or plant layer is removed. The level of carbon in the soil is what makes it great or poor. High-carbon soils can retain more water and drain well at the same time. It sounds impossible, but that is the magic of organic matter—good drainage and simultaneous water retention.

## People Do Care About Nature

People care what happens on this planet. The media and mainstream public are not so aware of this, but you should be. A lot of people care. I know: I speak to them all the time because of my work. There are many people who really care about frogs and rivers and oceans. There are countless people who love nature, are inspired by her, and value nature beyond any measure of money. I don't know why we are not more widely represented, but I do know that I can be a voice for nature. I refuse to be shy about how much I love trees and wildlife. If people think that's weird, I think they are weird. I love the Earth. In fact, how ridiculous is it to *not* love the Earth? And yet people will label you for doing so. They will call you a tree hugger or a radical. I think it is radical to cut down 95 percent of our forests, plow up all the grasslands, poison rivers until they are undrinkable, and kill people for cheaper oil. I don't think I'm a radical compared with the actions of my civilization. The weird thing is that most people in this civilization agree with me. They love trees and rivers and wildlife. We are all just caught up with the movement of the herd. We can see that the herd is not going the best way, but so many of us are not saying anything about it even though we care. When you speak up about your love for nature, you will be surprised at how many people feel the same way. But the important thing is to not just speak about it. You've got to do something, and that is the purpose of this book: to show you some great things you can do.

## The Caretaker

As humans we have a tremendous amount of power at our disposal. We can use our power in any direction. It's not necessarily positive or negative, creative or destructive; it is simply power. We can use our power to foster staggering abundance and diversity or we can use it to create a mass extinction. We can steer our power in many directions.

# Life Fountains

The human species can be caretakers for wildlife. We can enhance habitat more profoundly than any other animal. Beavers have their role, slowing creeks into ponds. Bees pollinate flowers, herbivores enrich the soil, birds spread seeds . . . every species has its role, its contribution. Ours has become lost to us, but it is one we can regain.

There was a time when humans enhanced habitat. They did it for themselves, as all animals do. It was beneficial to have large populations of wildlife for hunting. Forests were thinned heavily, grasslands were burned, fruit and nut trees encouraged. The results were abundant ecosystems that fed people and wildlife. Yet today we believe that food comes from farms and that farms are big open fields of one crop. The truth is much more complicated.

I believe there may have been more food in America before European contact than there is today. If you think that's crazy, consider these facts. Today there are 80 million cattle in the United States compared with an estimated 60 million wild bison 500 years ago. Pound for pound that's about the same, and those bison lived without fences, feed, antibiotics, or water inputs. They reproduced without artificial insemination. Those bison also lived alongside massive herds of elk, antelope, and deer, all of which fed on perennial grasses that never needed irrigation because their roots extended dozens of feet beneath the surface of the American prairie. There are stories of salmon filling rivers. Sturgeon spawns that fed ancient cities, billions of shellfish, and massive nut trees, like the ancient American chestnuts that covered the ground with nuts a foot deep. I'm not going to attempt to do the math, but I think the loss of fish, large herbivores, and giant nut trees has not been equally replaced with the calories that industrialized agriculture provides.

There is a fallacy that North America was once a wild continent where nature abounded because the people were simple and primitive. This is a lie propagated by a European culture that was blind to the decimated civilization they encountered. They could not see the managed landscape because it did not fit into their idea of farming. They did not see the millions of people who died in the wake of disease that spread faster than they did. There were cities in North and South America that were as big as any in the world at the time. Cahokia was a city in Ohio that had trade routes established from Nova Scotia to the Rocky Mountains.

# Trees of Power

If you have not read Charles Mann's book *1491*, I cannot recommend it highly enough. It shattered my beliefs about the American wilderness. This "pristine" land was managed by people. America was home to the largest forest gardens in the world, and to thriving cities. It was as populated as Asia or Europe at the time before contact. We were all lied to in school. Perhaps you heard the story of Cortés conquering the Aztec empire with a few dozen soldiers on horseback. The truth is, Cortés enlisted the help of the Aztec's rivals. Along with 200,000 native soldiers, Cortés fought the Aztec in one of the largest battles in ancient history. I don't know why we were lied to, why we were taught that Native American contributions were small and insignificant—that they were just simple nature lovers. Perhaps our teachers' teachers could not bear the thought of living on top of a decimated civilization that was able to thrive without degrading the environment. They felt too guilty, or maybe they were blinded by racism. Just remember, our version of history has been passed down to us by people who fully believed in slavery, oppression of women, and the white man's burden.

I'm sure there are lots of other things they left out. Archaeologists are discovering many of them today. It has been our belief that the Amazon was always a wilderness. We are now learning that it was home to a great civilization with extensive trade networks, enormous orchards, and one of the biggest cities of the ancient world, rivaling Egypt's. An entire civilization vanished from disease, leaving behind pottery shards, deep black soil, and giant food trees that still live today. These people knew how to stabilize tropical soils; they knew how to live in a way that enriched the land around them. They left behind some clues for us to learn how they did it.

Without a history of lies, we can see that people are able to thrive and nature can thrive at the same time. I actually believe that we can't thrive without vibrant nature. If you think our civilization is thriving, consider our rates of suicide, depression, and complaining. People in my country are among the richest and most miserable people in the world. Many of us hold deep-seated guilt about destroying oceans, rain forests, and the future, whether we admit it or not. No one feels good about the destruction of nature, not even the most narcissistic billionaire holed up in his tower. People care about nature; it is part of who we are. It tugs on all of our heartstrings.

# Life Fountains

We can be the caretakers. We have tremendous tools at our disposal, tools for digging, communication, and travel. We can share seeds from around the world and engage in some of the most remarkable breeding programs. Knowledge can be spread from one insightful person to millions of others. We can dig ponds and plant trees at a rate that surpasses any other civilization in the history of the world by a thousandfold. We can choose to use our powers to enhance diversity, deepen soils, and spread abundance. We have the tools. The time is ripe. People are starving for meaningful work. There is plenty to do. No time for self-pity or blame. It is time to plant trees.

Taking on the role of caretaker does not mean that ecosystems are just managed for wildlife. Caretaking means that we are producing food, medicine, fuel, and fiber from land that is enriched by our activity. For caretaking to actually work, it needs to be profitable. If it is not profitable, then very few people will engage in it. Time is precious. I cannot devote very much time to planting trees just for birds. In order for me to be caretaking, farming, and parenting at the same time, I stack functions. I coax into existence ecosystems that are beneficial to me and

This is still an ecosystem. It is a community of plants, birds, mammals, insects, and microorganisms, regardless of any opinions or judgments.

wildlife at the same time. It is in my best interests to stabilize and build soil, to have high populations of bats and insect-eating birds, and to maintain a highly diverse plant community. I pull carbon out of the sky, not because I am concerned about climate change, but because the trees I grow make money, pay my bills, and feed my family. It's important for caretaking to be in the best interests of everyone. Drawing a line on a map and designating a wildlife sanctuary only goes so far. We need land for growing food and for people's houses. We do not need bare-dirt monoculture farming and tract housing with sterile landscape shrubs next to the pristine, untouched nature preserve. We can have diverse farms that build soil and increase biodiversity. We can have neighborhoods filled with fruit and nut trees, berry bushes, medicinal flowering perennials, and gardens. Caretaking benefits us as much as it benefits wildlife. To draw a line between us and nature hurts everyone involved. Nature, the ecosystem, is everywhere, in every neighborhood and city that has a plant. You don't have to drive to a national park to see an ecosystem. There is one right outside your door. It may be dysfunctional, but it is there, waiting for your hands.

## To the Environmentalist

For lots of reasons, the environmental movement has become fairly unpopular today. The movement is generally viewed as a bunch of whiny liberals disconnected from where their food, cars, and lifestyle come from. Environmentalists are associated with holding up signs saying NO to everything. Opposition comes in hard and heavy because environmentalists are thought to be trying to stop industry and slow our precious economy. Fear of losing out on any profits has become justification for stomping out environmental justice. Like many topics, it has become polarized.

There is a better path forward. As environmentalists, we can view the whole movement completely differently. Right now, the movement is focused on reducing fossil fuels, creating buffer zones for wildlife, and saving endangered species. I get it. It's outrageous how our culture treats nature with no respect, no reverence. It's disgusting to watch millions of acres gobbled up by bucket wheel excavators on the tar sands. It's utterly depressing to hear about another million-barrel oil spill in the ocean.

It's hard to take when we see the trajectory of climate change. There is a saying in the environmental movement: "If you're not outraged, you're not paying attention."

I understand that completely. I have chained myself to government facilities and gone to jail; I've screamed my head off at dozens of organized protests. I understand the frustration and the need to stop the destruction. I also now understand that anger is a very ineffective tool.

When you show someone anger, they will instantly become defensive and then, often, offensive. You won't convince anyone of anything by yelling at them.

The best way to create change is to create alternative options that are so much more appealing than the status quo. For example, let's say you are concerned about palm oil plantations destroying rain forests in Southeast Asia. You could tell everyone you can to stop buying palm oil; you can make a video about it, write a book describing the horrors of deforestation and the loss of orangutans. You can shout and shout about it, but at the end of the day, people are often going to buy what is on the shelf at the store. If you wanted to get people to buy less palm oil, then at some point you have to offer an alternative, and it should be better than palm oil. Telling people all about the hazelnuts you grow and process into oil is a lot more inspiring than trying to make them feel guilty about buying palm oil. You will get a lot more traction by offering something new or creating positive choices for folks to make. If you just say no all the time, then you are actually a negative force. To create positive change, we have to be a creative, positive force. It is much harder than traditional protests. It takes a lot of energy, knowledge, inspiration, and faith. That faith comes from understanding our sphere of influence.

## Sphere of Influence

The problems of the Earth are overwhelming. Looking around I can see my government engaged in a never-ending war, tar sands stretching across western Canada, deforestation of the tropics, the demise of polar bears, climate change, nuclear proliferation, drones killing families, child soldiers, dolphins caught in nets, hydraulic fracking, the poisoning of the Niger River, the melting of the Arctic, hemlock woolly adelgids, mountaintop removal, and a wasteful culture accepting of all this irreverent destruction. It can get a person down. It is overwhelming unless we work within our own personal sphere of influence and trust others to do the same.

I have been inspired to plant trees and create a new kind of agriculture, one that builds soils, slows climate change, feeds people, and increases biodiversity. I cannot work on that and work on nukes or the oceans or prisoner rights. I don't ignore those issues, but I would be less effective if I thought I needed to fix everything. I trust that there are people who care deeply about the Appalachian Mountains and are working to stop mountaintop removal. I trust that there are people who are working on the rights of prisoners and others working on the acidification of the oceans.

My sphere of influence is determined by my inspiration and my reach. I speak to people who might plant trees. I don't waste my time on congresspeople. I trust that someone else in the environmental movement is following their passion. I am grateful for people who are out there talking to Congress, and those standing up and blocking pipelines, and the ones in the South Seas intercepting illegal whaling ships; I am grateful for all the work being done out there by the children of the Earth. I wish I could join them all, but I know that I would be less effective. Following my own inspiration, I trust that others will follow theirs.

We can create real changes if we are not overwhelmed. I can work efficiently and productively within my sphere. The ripple effects of my work may carry much farther than I imagined, but I cannot get caught up in trying to save the whole world. I stay on point and plant tree after tree, all the while inspiring others. My sphere of influence may have started small, but by staying committed to my inspiration, it spreads and grows just like my trees do. Working from a place of inspiration is very different from working out of anger.

It is impossible to see the effects of your actions. They ripple out and out. Yellowstone National Park has a dark history of wildlife eradication. When the park was first created by Theodore Roosevelt, retired soldiers were hired to kill all the predators. Among many other effects, herbivore populations exploded. The land became highly imbalanced and the ecosystems grew fragile. In recent decades, though, wolves were reintroduced into Yellowstone. Researchers watched eagerly to see what effect this would have on herbivore populations. What they saw went far beyond a balanced population of predators and prey. Elk and deer changed their habits. They no longer freely browsed anywhere. Instead they became elusive and stayed out of the open. Vegetation along rivers and streams skyrocketed. Erosion was greatly reduced and the course of rivers literally altered. If adding wolves to an ecosystem can change the course of a river, what unforeseen effects will your inspired actions create?

I once heard a great quote from the author/caretaker Derrick Jensen when he was giving an interview about the environment. It went something like this: "It's as if someone is at the ER and they have some stab wounds and the doctors are trying to patch them up, but while they are sewing stitches, the attacker is still stabbing them." This is what it feels like to stand up for nature, to be an environmentalist. You are like the doctor, stitching wounds while more are being made. I'm grateful for the people who are stitching up the wounds—we need them. I want to stop the attacker. I want the attacker to turn their head away from their stabbing because they see something beautiful out of the corner of their eye. I want the population to become aware of positive, profitable alternatives. I want the attacker to put down the knife, not because I scared them, but because I inspired them. Punishing and berating the attacker may stop them for a moment, but sooner or later they will do it again. To create real change, we've got to *create* real change.

## Wreck the World Slower

A lot of the environmental movement has become associated with things that are intended to slow down destruction. Efficient lightbulbs, electric cars, solar panels, recycled paper, shorter showers, and veganism are all efforts to slow down ecological collapse. I don't want to wreck the world slower; I don't want to wreck it at all. Follow your inspirations, not your fears. We environmentalists can be overwhelmed by the destruction we see, but we can move from a place of creation. We can shine a light so bright that even the blind money-chasers of Wall Street will see it. The only way to do this is through true inspiration, by following passions and dreams. It is necessary to understand that you are not bad for the Earth. You are not a mistake of creation or a disease. You are simply energy. You can create some of the most abundant and diverse habitat ever seen. You have tools that can make you more effective than your ancestors. You can tiptoe around hoping not to hurt anything, or you can educate yourself, step in, and make the world better.

Perhaps you have heard the phrase *Leave no trace* or *Take only pictures*. This epitomizes thinking in the environmental movement—that nature is fragile and needs protection. The truth is, we are all part of nature. It cannot be separated by lines on a map. Our involvement in nature can enhance or hinder biodiversity. It is a matter of making educated choices. Instead of trying to have as little impact as possible, I want to have a huge impact. I want to leave behind millions of trees, a bunch of ponds, enriched soil, and wild stories. If the goal of environmentalists is to have no impact, then the best thing they could do would be to die—that would have the least impact. Even in death, though, their body will feed the soil. We are having an impact no matter what. Make it one that you are proud of. If you are not sure how to do that, then listen closely to the world. Look at nature everywhere you go and see what is happening. Learn to identify plants and trees so that you can actually read the story being told by the world, not the news. If you seek a path forward, you will find it. Nature is talking loudly all the time; you have just been domesticated and conditioned not to hear her. Learn as much as you can about trees, plants, animals, fungi, and ecosystems, and you will see so many places where you can step in and become a powerful positive force. In fact, once you are able to see what is happening in ecosystems, you will find yourself with nowhere near enough time to act out all your inspirations.

## Taking from Nature

We have no choice but to take from nature. We must eat and drink. Many people are so disconnected from their food that they don't realize how much taking there is. Every day, it is a lot. In my early 20s I became obsessed with nature and wilderness survival. By living simply with primitive skills, I formed a connection to my food. I took every drop of it myself and saw just how much I took. I killed constantly to feed myself the bodies of plants and animals. At first I felt guilty and tried to take as little as possible. I had an attitude of apology. In time I learned to develop an attitude of gratitude instead. We are part of this system of carbon eating carbon, life transforming itself through the endless chase of animal to soil to plant to animal to soil. During my first full survival campout, I learned that my body was directly powered by what I ate. If I ate nothing, I had no energy. If I ate frogs and cattail roots, then I was powered by that energy. I literally take them inside me.

What we perceive as taking from nature is not actually taking. Energy is neither created nor destroyed. You are taking life to have life. Something is killed for you to have energy. When I kill a deer and eat it, I have that energy in my body. What I do with that energy matters more to me when I remember this. Am I taking life so that I can lie around and feel sorry for myself, or am I taking life so that I can spin webs of beauty, spreading kindness and seeds? I don't remember this all the time, but I am grateful whenever I do. It's important to give back and feel good about what you take. Don't be hard on yourself; treat yourself with the same kindness that you would wish on anyone else, and go out there and do some good work. If you don't know what that work is yet, then listen very closely when you interact with the world. Sooner or later, if you're listening, you'll find the work that is meant for you. We are not a mistake of creation. We have the same value and right to be here as everything else. It's okay for us to take from nature; it is how we can give back.

I was once talking to a woman who was telling me about the cherry tree in her yard. It was dripping with fruit. She felt guilty taking from the tree; she thought she shouldn't take too much. I asked her, "Why do you think that tree made the fruit so good? It wants you to take it and spit the seeds out all over the place. If you ever feel guilty about collecting food from nature, then find a way to give back." Some years I collect a ridiculous amount of hickory nuts. I realize there will be less for the squirrels (I harvest a lot). I can offset this by planting and establishing just a couple of hickory trees. The trees I plant and establish will live beyond my time and feed countless squirrels and other wildlife. You don't have to feel guilty; you can be grateful. You can plant trees to be sure that your presence on Earth will lead to wonderful events. You are designed to take from nature. It is the same for all creatures on this planet. Do so consciously, and you cannot help but create benefits for wildlife.

## Having Energy to Do the Work

Environmentalists can burn out pretty fast. Traditional environmental activist work is tiring and largely unrewarding. How long can you carry on protests when it seems like no one is listening? Grant writing and lobbying involve time indoors in front of screens. This is pretty tough on the soul of a nature lover. It is important for environmentalists to directly connect with what they are fighting for. Spend time in the woods, in the fields, or the ocean. Know nature, feel nature, and return to the fight with a fire in your heart that cannot be put out. Return to nature constantly for rejuvenation of spirit, for new ideas, and for a healthy mind and body. You cannot just sit in an office and be an environmentalist. You have to also be in the environment. To be truly inspired, you may have to actually get outside!

Doing the work takes a lot of energy. Planting, weeding, pruning, felling, digging all take energy. Feeling lazy won't go very far; being tired won't leave a wake of inspiration, and it sure as heck won't bring back the chestnut. Understand that you do not have a finite amount of energy. Watching TV and lying on the couch all day will not save energy for you to use later. A human is more like a wellspring. Taking water from a well causes new fresh water to come in. Leaving water in a well leads to stagnation. It is the same with your energy. The more you use, the more will flow freely into your body.

Stretch your body, breathe deeply, exercise, awaken your senses and run. You will not wear yourself out by doing things. You will run yourself down by doing nothing. Wellsprings of energy exist in you. The more you use them, the more they will flow. You are part of creation, a powerful and mysterious force. Don't diminish that by thinking you are just a rain barrel with a limited amount of water. You are the wellspring. You don't need to save your energy for later. Use it now, and you will have even more. People who run ultra-marathons (100-or-more-mile races) run all the time. The more they run, the more energy they have. They don't sit around for a month saving energy until the day of the big race. It is the same with the Earth caretaker. I can work 10 times faster than I did 10 years ago. My movements are efficient far beyond my past. I am filled with more and more energy the more I do.

Sometimes I need to rest, and I do. There are also times when I feel tired and lazy and I don't want to. The best thing to do when that happens is to get up and stretch. Start moving, start breathing deeply. Yoga, Wim Hof techniques, running, meditations, martial arts, cold-water immersion, and extremely deep breathing are all methods for increasing energy. Get over the inertia, and the energy will start to flow when you move. The more you are aware of the energy in your body, the more it will be activated.

With increased energy, we are more effective caretakers.

## Plants Are Sentient Beings

I still remember the first time I realized that plants were living, sentient beings. It was at night in a city park in Boston. I was 19 and becoming awakened to the aliveness and beauty of the world around me for the first time. I remember sitting on the grass. It was a warm night. I was staring at the stalk and seed head of a narrow-leaf plantain (at the time, I had no idea what this plant was called). It is a common weed of lawns, but on this night it was clearly alive. The way that it arched is burned in my mind. That image is with me today, full of power. I could feel that this plant felt me; it was aware of my presence. This experience is beyond description. I have known without a doubt ever since that plants are alive as much as I am.

Scientists are proving that plants are cognizant, sentient beings. Monica Gagliano has conducted incredible experiments showing that plants have a sense of hearing and memory. She is one of the leaders in the newly developing field of plant cognition. In one experiment a plant is grown in a pot shaped like an upside-down Y. The roots can grow in either direction of the Y. Without any inputs, they grow in both directions equally. However, when a tape recorder of rushing water is placed next to one side, most of the roots grow to that side. That is a tape recorder, not actual water. They are sensing the vibrations of sound and responding. I can define sensing vibrations of sound as hearing.

In another experiment plants are placed in a dark room. A fan is turned on from one direction. After the fan is turned on, a grow light is turned on from the same side as the fan. Naturally, the plants turn to face the grow light. This is done over and over until one day only the fan is turned on. The plants continue to turn in that direction with only the stimulus of the fan, no light. They have a memory of the light coming after the fan. They don't have ears, but they can hear. They don't have a brain, but they can remember. There is more than one way to perceive the Universe than just through the human mind.

Personally, I don't need to prove it; I already know that plants are creatures just as I am. If you don't know it, that's okay. Your perspectives are your choice. Whatever thoughts you choose are chosen by other thoughts. You have no idea where any of your thoughts come from. They rise from a dark and mysterious space. They come into your mind and say things. Maybe you believe them, maybe you don't, but who is the one doing the believing or disbelieving? Just more thoughts. Your mind will say untrue things. How can you believe every thought? You are just guessing at which ones are true. If you listen close enough, your thoughts will contradict one another. You don't have to believe anything. Reality is stranger than any scientist could handle. The Universe is infinite and weird. We are deep inside a dream, far inside. As you exist, breathe deep and appreciate the wondrous world around you, all the while feeling your existence. Do not shy away from the awareness within your own body. It is a constant place for you to come back to whenever things get crazy. You can always find the center if you let everything go for a moment and feel your actual center. Don't think about it; just feel what's deep in there. Then look out: You can walk around knowing that there is a presence inside you. A presence that is strong and unwavering. Once you find this presence inside your body, you can always find it again, whenever you remember. It's there; if your mind would stop talking, you could notice it.

Looking at that plantain on that night in Boston without any thoughts, names, or opinions is how I was able to really *see* it. We all have the ability to breathe deep and be quiet. Just be quiet and listen. Plants are created by soil, sunlight, and water. Nobody really knows what those things are. They are ancient forces beyond the scope of thoughts in your human language. Plants eat sunlight and stretch themselves toward the stars. They are here because creation wants them here. They have their own ways of expressing life, of communicating with creation. Listening to plants will enliven your senses. It requires patience, silence, and openness.

Scientific studies have shown that plants respond to being talked to. For those of us who already talk to plants, we don't need to see the studies. That is for the doubting mind. Doubts are choices, as all of our thoughts are. Faith in the presence of life force everywhere in everything is all I need. I can see that plants are feeling life. I know some of you can, too. Our society is ruled by the cynical mind, but you don't have to be. You can choose to believe that creation is more than what you know. The Universe is filled with exploding stars and ancient rain forests. Our view is limited, as we peer out from two eyes at such an ancient vast mystery.

I believe that rocks, stars, planets, plants, water, animals, and myself are all filled with the power of creation. We are of the infinite and always will be. It is not that humans are conscious and nothing else is. There are no lines to draw, separating this from life and that from death. We see trees growing out of rotting logs filled with fungal threads, or mycelia. You don't know how to speak to most life-forms in the Universe. It is not because they are not alive; it's because you don't know everyone's language.

Plants have lived for millions of years without people. What do you think they were doing that whole time—waiting for us so they could be used? They are here and always have been because they are an expression of creation. They are composed of living cells.

It takes an extremely egocentric view to believe that all the creative intelligence of the Universe is held only in human brains. Yet that is the view of our culture. It is considered absurd to believe that stones, stars, plants, and animals have as much sentience as we do. Just because life is experienced differently in different forms does not make it any less real. We have access to five senses. How many others do we not know of? What do spiders perceive? What do stars feel as they burn in deep space? These are not questions our minds can answer because they lie outside our minds. The Universe is more vast and complex than any scientist can show in an experiment.

## Finding Mother Trees

Being able to identify trees is as useful to me as reading. I can identify species with a glance at bark or silhouettes while driving down the road. It may sound hard to do if you're just beginning to learn identification. The truth is, learning to read is also hard to do—just ask any five-year-old. If you practice and are patient, though, identifying plants will become as easy as reading.

It doesn't take a formal education to notice that this crab apple has exceptional qualities. I cannot imagine passing by a tree like this and not collecting seeds or cuttings.

Once you can identify trees, they will reveal themselves to you everywhere. I find outstanding individual trees behind office parks, along highways, in yards—just about anywhere.

It has become an ingrained habit for me to look at trees everywhere I go. I can map out areas for many miles from my house based on tree species. If I want to collect a type of seed, I know where I can check for it in several locations because I have been paying attention for years. I have so many special spots all over the state.

I used to think that special trees worthy of being a named variety were extremely rare. I now think I could find some anywhere I lived in the world that has trees. Keep your eyes open, pay attention, learn to identify trees, and you will find treasures.

About eight years ago I noticed some ducks and geese hanging out under a mulberry tree on the side of the road. I stopped the car. That is part of the key: You have to be willing to stop and look at things up close. When I got out of the car, I was blown away by the beauty of this tree. It had a perfect form with thousands of branches covered in berries of various stages of ripeness. The ripest were jet black, their flavor unsurpassed by any berry I had tasted. This tree is one in a million; I've revisited it at least a hundred times. It bears fruit every year, with the first berries ripening in June. There is a steady supply of ripening fruit every day until the beginning of fall. I named this tree the Everloving Mulberry. It was just growing wild on the side of the road. There are many more special trees dotting the landscape. Most will never be discovered by a population that has become illiterate of the natural world. How can we find these treasures if we can't even identify trees? Who will find the next amazing apple or the next shagbark hickory with golf-ball-sized nuts? It will be the people who are looking around everywhere with rapt attention to the abundant and varied plant world that surrounds us

## Learning to Identify Trees

This is an invaluable skill, both economically and for personal enjoyment. The actual learning takes patience and practice, but it gets much easier the more you work at it. In the beginning everything will look similar. As time goes on your eyes become trained to nuances that slowly turn into brightly lit signs. I remember struggling to tell the difference between Norway maple and sugar maple—they looked so much alike. I kept working on them, looking at them every chance I got. Today I can differentiate Norway and sugar maple with a glance any time of year, any part of the tree. The differences are huge, and the two trees about as easy to tell apart as a pickup truck and a car. My eyes became trained. It happened by looking and looking again. The following paragraphs are the techniques I used to teach myself how to identify trees.

### Drawing

You don't have to be any kind of artist. Choose a leaf or twig with buds to sketch. Don't just look at it and copy it. What you want to do is stare at the leaf or twig for a minute or so and then set it down. Draw from memory. You will quickly find the gaps in your memory. Look at the leaf again and repeat the process. You can do this as many times as you like. It does not matter if you can identify it or not. What counts is that its details become familiar to you. Once it is familiar and you know it, sooner or later you will come across it in a book and it will jump out at you. You will know it very well after that and never forget it.

## Twig/Leaf Board

I've made lots of these and they can be quite beautiful. Take a piece of cardboard or poster board. Attach winter twigs or pressed leaves of different species. When they are next to each other, it's amazing how their differences will stand out. You can write the names under the twigs when you figure out what they are. I used to hang these boards in front of my toilet so that I often had a chance to study twigs.

## Books

A good field guide is essential. It should be one that you enjoy looking at. Some will have pictures and some illustrations. Make sure that you like the feel of the book. You want to flip through the book all the time—on the toilet, in line at the store, whenever you think of it. It's nice if it fits in a back pocket. Flip through the book often enough so that when you see something in the wild, you might think, *I remember that from the book*, and then go find it in there. Tree ID books I personally love are the *Golden Guide to Trees of North America* and every book by William Harlow.

## Learning Barks

Identifying trees by bark is the most difficult method, but incredibly useful. When walking through a forest, it's almost impossible to see the leaves or buds on many trees. When you can identify bark, though, you can walk through a woodlot and recognize what is going on without looking like a tourist staring at skyscrapers. The way I learned barks is by first learning twigs and leaves. When I identified a leaf, such as red oak, I would then look at the bark—every time. Red oak leaf or twig, look at the bark. Touch the bark. If you recognize a tree, look at the bark. Doing this enough times will imprint the bark's patterns in your mind. Over time, barks will look very distinct from one another and it will be easy to tell them apart.

Tree barks change dramatically with age and from one individual to the next. There are many variations within each species. Through repetition, you will create familiarity, and before you know it you will recognize every trunk in the woods.

The texture of barks varies greatly. It's important to rub your hand across the trunk. Rubbing your hand over a red maple trunk and then a sugar maple trunk are very different experiences. Try it.

Being able to recognize trees by their bark, twigs, leaves, or silhouettes is like having a superpower. You will be able to read stories in the landscape. You will find treasures. You can even make a lot of money if that's what you want to do. I've made tens of thousands of dollars by being able to identify trees. I have collected uncountable amounts of fruit, nuts, and seeds that have become the backbone of my business. If I travel through an area, I almost always find something to gather because I can read the signs. Imagine if you were trying to find a street in a city but could not read any signs. That's what it's like for people walking through neighborhoods and forests who can't identify trees. Educate yourself. It's a shame that we weren't taught this skill in school, but there's nothing to do about it except change it. Learning to identify trees is not nearly as hard as many other skills. There really aren't that many species to learn in any given bioregion. Learning a dozen trees will get you well on your way.

## Breeding

If we collect seeds and grow them out, we are participating in the evolution of a plant. We are breeding through any act of growing seeds. We can do this

The twigs of different tree species are very distinct when looked at closely. Here is shagbark hickory on the left and American beech on the right.

unconsciously by collecting whatever seeds we can reach from any old tree. Consciously, we can steer progression by collecting from specific mother trees, and we can take it a step further by collecting from specific father and mother trees.

Universities conduct most of the formal plant breeding programs today, but that was not always the case and it certainly doesn't need to be today. People who interact with plants are the most qualified breeders. Regular growers can identify plants with exceptional qualities and collect their seed. It really is as simple as that.

It is absolutely true that trees will not come true from seed. That is, each seedling will be different from its parent, just like your kids will be different from you. Some traits will be carried through to the next generation. If two people with red hair have a kid, there's a better chance the kid will have red hair than if one of the parents has black hair. If you're selecting seed from trees with straight, vigorous timber form, there's a good chance the kids will also exhibit that. It's harder for fruit flavor to come through generations, but it does happen.

If you are finding special trees in the wild or in your plantings, it is completely within your ability to grow seeds out. You don't need to rely on institutions to do this work. Many of the best varieties of plants discovered were grown by regular folks.

The more seedlings you grow out, the better your chances of finding exceptional performers. You can plant trees densely to find these special "genius" trees. You can take this to many levels. For example, a 4-by-20-foot bed can fit thousands of apple seedlings in it. During their first year, they can be sprayed with fireblight (a bacterial disease) and you'll be left with all the resistant ones. Another, less intensive example would be to plant chestnut trees 6 feet apart instead of the standard 40 feet. Over time the disease-resistant best producers will reveal themselves. You don't have to understand cross-pollination or have a degree to breed plants. All that is needed is to plant a lot (starting with good parents) and then identify the best offspring.

This is how breeding has happened for thousands of years. It is how people developed virtually all the domesticated crops we eat today. Modern agronomists still cannot duplicate the work that indigenous people performed to transform the wild grass known as teosinte into our modern, domesticated corn. One of the greatest plant breeders of all time was Luther Burbank, a man with no formal education, who developed more plant varieties than any institution. He developed the russet potato along with 800 other varieties of fruits, nuts, and vegetables. Another grower/breeder, Ephraim Bull, discovered Concord grapes by planting out 20,000 grape seedlings.

Plant a lot of seedlings and you, too, will find some special ones.

Controlled crosses involve bagging flowers just before bloom and hand-pollinating them with pollen from selected plants. Seed from these fruits are then grown out. It's a step above just collecting seeds from a good mother plant. In controlled cross-pollination, you are selecting the mother and father plants.

# Hybridizing

Hybrids occur naturally in nature. Oaks regularly hybridize in the wild, as do hickories, butternuts, apples, and many other trees profiled in this book. A hybrid is where pollination occurs between two distinct species. The resulting offspring is a hybrid. I only mention this because many people immediately associate hybrids with something highly unnatural. Hybrid trees are nothing like genetically engineered plants. They are merely created by mixing the pollen of one species with another. Generally, hybrids only occur within a genus. For example, bitternut hickory could be crossed with shagbark hickory, but it's not possible to cross bitternut hickory with apple using traditional breeding techniques.

Before I explain the benefits of hybrids, it's important to understand where different species may have come from. Let's take chestnuts, for example. There are several species of chestnut found around the world. There are American chestnuts (*Castanea dentata*), European chestnuts (*C. sativa*), Japanese chestnuts (*C. crenata*), Chinese chestnuts (*C. mollissima*), and many others. All of these trees have been isolated from one another for many thousands of years. They have developed their own adaptations to their respective environments. At the same time they all have a lot in common. All the chestnut species make a chestnut—a very specific seed. How did all these different species make the same kind of nut growing isolated from one another? The answer is obvious to me: They all came from the same place.

The entire *Castanea* genus shares a common origin. I find it hard to believe that each species of chestnut sprang up on its own. I know the world is mysterious, but this seems a little too unlikely for me. An ancient chestnut species spread in different directions. Over time they became isolated from one another through the formation of mountain ranges, glaciers, and continental drift. Each species moved on to its own specific habitat and learned different adaptations.

Today diseases and insects are spread around the world at an accelerated rate. All of the different locations that individual species adapted to have changed. The world is not static. One place never remains the same. Species either adapt or vanish. Chestnut trees have learned a lot about different bugs, fungi, viruses, and climate extremes. Their knowledge is scattered among individual species around the world, from Maine to Japan. We can bring this knowledge together by bringing chestnuts back to one another. The trees readily hybridize if they are planted near each other. Chestnuts all started out together from one center of origin. We have the opportunity to bring them back together so they can share the best traits and adapt to a rapidly changing world.

Japanese chestnuts have excellent resistance to chestnut blight; American chestnuts are extremely cold-hardy. By allowing the pollen of one tree to find the pollen of another, we can find trees that possess both of these qualities. The American chestnut would not need to adapt to the blight if it were not for globalization. The best hope for chestnuts is for them to come back together after their diaspora. We are the way for them to find one another's knowledge.

I believe it is my job to partner with chestnut trees. I offer them steps forward in evolution by bringing them together. If you are a native-plant fanatic, then you probably don't agree with this approach. It's important to remember that native plants are native to a place, but places change. Soils, climate, insects, animals all change. If we deny trees the ability to change as fast as the world's climate does, it leaves them with less chance of surviving this volatile period of Earth's history. We are facing a period of major species extinction. But we can give species the best chance they have using their own tools of cross-pollination. Let them share with one another what they have learned as they lived in different parts of the world. The world is coming together, so there is no reason to leave the trees out of this conglomeration of knowledge.

The time for reunion is here. I believe hybrids are the way forward in a world demanding rapid adaptation.

## Money Grows on Trees

Contrary to the popular saying, money *does* grow on trees. It simply needs to be converted, like exchanging different currencies when you cross an international border. There is value just falling on the ground. I have made around $100 an hour collecting wild pears. That's pretty good money even for someone with a degree and a suit, but I have made even more than that by realizing that money grows on trees. There have been times I have been able to collect thousands of dollars' worth of seeds in a day. I see grafted mulberry trees growing on my farm; each foot of each branch is worth $5. I know that money grows on trees. Money just means value; it means one thing is worth another in a barter. If you need money and you get it from trees, you will have a whole new appreciation for the abundance that rains down from their canopies.

I have tied my livelihood to trees. I completely rely on them to pay the bills. This may sound risky to some, but I believe it is very safe. Not that there is one tree I would really bet everything on; instead, I have a diversified portfolio. Instead of investing in McDonald's and ExxonMobil or Lockheed Martin, I am invested in chestnut, hazel, persimmon, mulberry, apple, pear, and 50 other species. At least a few of them will have a phenomenal return each year. My investment costs are low or nonexistent, and the returns are over a hundredfold and ever increasing. I also feel great about what I am invested in. No one has to be killed, no river poisoned, no rain forest cut down for my investments to pay me big returns.

When we make money from trees, we are literally invested in them. You can't help but care a little more about what happens to trees when your livelihood is tied to them. I have heard from lots of people that they hate their jobs or they don't believe in the mission of the company they work for. There is always a way out, always other options. Trees offer alternative forms of income to creative, hardworking folks. Some people are in tough situations; they work multiple jobs and take care of small children. Working with trees on the side can provide extra income. A lot of the work I do with trees happens while I am taking care of my three young kids. Everyone's situation is different, but there is an astounding array of options for gathering money from outstretched branches in every city and rural area.

I have included a section on the commercial possibilities of each tree in this book for those looking to supplement or replace their current stream of income.

CHAPTER TWO

# Planting

P lanting trees is a powerful act. We can influence the lives of people and wildlife for centuries with a successful planting. I hope that this chapter can help you get your plants off to a strong start. Planting trees is as much an art as it is science. Techniques can be constantly changing as we learn from nature's responses to our actions. There is no one right or wrong way to plant trees. It is not like starting up a machine; plants are infinitely variable, but they all hold one thing in common: They want to live. They will find a way even if you mess things up a little.

When I am planting trees, I often have in the background of my mind a vision of what it will be like after I am gone. I think about huge spreading trees that are feeding generations of squirrels, raccoons, opossums, grouse, turkeys, white-footed mice, songbirds, deer, and kids. I imagine these trees being climbed into and making their own seeds. I imagine them being magnetic centers for wildlife and people when fruit and nuts are falling everywhere. But for now, the tree is small. It is only one or two years old. There are so many practical considerations for a successful planting. Here are some key points that I hope may help you establish trees.

# Water

Water for trees is the same as it is for us: They want some, but not too much. Imagine if someone forced you to drink cup after cup of water—eventually you'd get sick. And of course without enough water you would feel terrible and could even die.

The best way to balance water for trees is to have high amounts of organic matter in the soil. Remember, organic matter holds four times its weight in water and is full of capillaries that drain and disperse moisture. Adding a thick layer of mulch around young trees will not only add lots of organic matter, but also trap moisture underneath for a long time. The thicker the mulch, the more fully this is achieved. In my area, a 6- to 12-inch layer of wood chips will retain moisture under it the entire growing season in most years.

I usually don't water trees at the time of planting. I generally just plant bareroot trees in the fall or early spring while they are dormant. At these times of year, the ground is already moist and the trees have very little need for water. They do not have leaves on them that are transpiring moisture. I do my best to mulch each tree and then check on them once the growing season is under way. If it has not rained in a long time, then I will pull back the mulch and use my finger to determine if the soil is dry. Only in stretches of drought do I even think of watering my permanently planted trees, and even then I only water trees that are one or two years old. Please keep in mind that I live in the northeastern US; you may be in a very different climate. I personally don't like the idea of irrigating my trees. I want their roots to dive down deep, spread far, and not be focused around irrigation lines.

If you do supply water to your trees, understand that there are key times to do so. When trees are most actively growing is in early summer. They will be making most of their growth during this time and using a lot of water. Toward the end of summer and early fall, they should be hardening off (turning tender growth into wood). Supplying irrigation or fertilizer while trees are hardening off can be detrimental. They often will keep growing only to be burned by the first frost. Water in moderation at the right time is key. I believe it's best achieved through working with the soil.

When you're moving bareroot trees from place to place, never let their roots dry out. Think of it as moving a fish from one pond to another.

## Planting

Their roots should always be covered with dirt or mulch, or they can be soaking in a bucket of water. When I am planting in the field, I like to keep my trees soaking in a bucket while I am digging holes. As each tree goes into the hole, its roots are coated in moisture. Moist roots go into moist ground, which is then covered in thick mulch while the tree is dormant.

## Heeling In

If you are not able to plant bareroot trees right away, then they should be heeled in. This is a very easy way to store them. Simply dig a hole or trench in a shady spot, lay the roots in, and pack it with soil. Make sure that the roots are completely covered and air pockets are removed by firming the soil. Dozens of trees can be bundled together in the same hole. They can stay heeled in until they begin to break dormancy. However, it is best to plant as soon as possible. The longer your trees are in the ground during the dormant season, the more time they will have to establish themselves without being stressed.

Bareroot trees being heeled in to a trench until I'm ready to plant. These trees are persimmons and naturally have black roots.

## Planting

Spread roots out, don't wrap or turn them to fit them in a small hole. Bent or J-roots can strangle a tree's trunk in the long run. It is better to prune roots or to make the hole bigger. It is okay to prune roots; it's not that different from pruning branches, just try not to take too much off. You can also dig out a small trench for long individual roots if you do not want to cut them back.

Plant trees at the same depth as or slightly deeper than they were in the nursery. Look for a line of changing color to see where they were growing before. You can plant most trees and shrubs deeper than they were grown in the nursery. Exceptions include oak, hickory, chestnut, maple, pine, and anything else that would never root from a cutting.

It is a good idea to bury graft unions, especially in northern regions. This protects the graft union from cold damage. It also allows the scion to root over time. Many folks advise against burying graft unions because the scion will root, therefore negating the effects of a dwarfing rootstock. If your trees are not on dwarf rootstocks, then it doesn't matter. Let them root and be strong.

Once the tree is in the ground, firm the soil with hands or feet. Take care to remove any possible air pockets and help the soil to settle. I don't stomp the ground, just press it firm.

## Deer

Protecting trees from deer is essential. They can keep trees forever pruned at a couple of feet tall. They will usually revisit the same trees over and over.

A section of fence 5 feet high, cut into an 8- to 10-foot length and bent into a hoop, serves well to prevent deer damage on individual trees. I place one sturdy stake on the inside of the hoop of fencing. The stake is woven through a couple of holes in the fencing. This way it can slide up and down the stake without too much difficulty. This allows you to lift up the cage if you need to pull weeds or put on a mouse guard. The stake should be placed on the side of the prevailing winds; this allows one stake to work by itself.

## Planting

Tree tubes are another option. There are many advantages and disadvantages to them. They require excellent staking. They also attract paper wasp nests. Trees grow very fast out of the tubes, almost too fast. They are so whippy that they can flop over without a stake. Slicing tubes vertically once the trees are taller than the tube allows trees to bend in the wind and strengthens their trunks. I leave tubes on after the tree is taller to prevent deer from rubbing antlers in the fall. Tree tubes in northern climates need to be vented to allow cold air in. This helps trees know that winter is coming—they need to harden off. Still, I have found that with vented tubes, trees still have a difficult time hardening off for winter.

For larger plantings it often makes sense to just fence the whole area. Standard deer fence is 7 or 8 feet high. Deer can jump higher if they really want to. You can get away with a shorter fence by understanding where the deer travel on a property. I have a 4-foot fence that has not had any deer jump over it in the three years it's been up, and a separate 6-foot fence that's not been jumped in seven years. I do have a lot of deer here. I made sure when I put up the fence that I did not block off any major corridors of travel. The deer are not looking for a hard time; they like to move freely and casually through the world as they browse. I think of deer fencing more as funneling them around an area than as putting up a wall.

If you totally cut them off, they will test a fence repeatedly and endlessly, especially if there is a bunch of fragrant food in there like apples or persimmons.

You can create your own corridors for deer travel by simply mowing a trail around the outskirts of a fence. You'll be most effective by knowing where they are going and why, but simply mowing the trail around the fence will help a ton.

Electric fencing tends to fail at some point. It needs weeds not to interfere and snow not to weigh down the wires. Deer will repeatedly test electric fences in my experience and often break through them.

An 8-foot woven wire fence is the safest bet if you have high deer pressure.

# Voles

These little rodents can be a major problem in old hay fields or in thick grass. During the winter, voles chew young tender bark off fruit and nut trees—both the trunks and the roots. While deer will keep trees pruned, voles can kill them completely. To protect from voles, I use white spiraled plastic tree wraps. They are inexpensive and easy to put on and take off. All rodent protectors should be taken off during the summer. Leaving them on too long can kill the tree by girdling it (even the white spiral wraps can do this). Also, taking wraps off allows me to check for borers or other insect damage on the trunk. Aluminum foil is another possible vole stopper.

Raptor roosts, such as dead trees or power-line poles, help in controlling vole populations. So do hedgerows that provide cover for predators like foxes and feral cats.

The best protection against voles that I have found is tight mowing in the fall. I will let weeds and wildflowers go crazy all summer, but if the grass is kept short in the fall, the voles will leave. They only feed on bark in the winter when green plants are scarce. These creatures are terrified of hawks, owls, foxes, cats, and coyotes. Their best defense is to hide in thick grass or under snow. You can't stop the snow, but you can keep the grass short in the fall.

# Soil

Soil is the most important aspect of any planting. Soil will determine the performance of your plants more than any other factor. Caring for soil is like religion: Everyone thinks their way is the right way. Just as there is more than one way to connect with the Universe, there is more than one way to approach caring for soil. Whatever you read and hear about soil, pick and choose what you want. You don't have to believe what soil scientists or other gardeners have to say. Over time you will create your own set of evolving practices and beliefs about what is happening underground.

### Life in the Soil

*A complex ecosystem bustling with life* is an accurate description of a handful of dirt. There is more occurring in the soil than we will ever grasp. We have recently discovered the tip of the iceberg with electron microscopes. Once you go down the rabbit hole, there is no end. Just to give you an idea of how active the soil is, consider root hairs. Trees make root hairs off every root. Some of these hairs will develop into a full-sized root, but many simply remain hairs. They have a life span of only a few seconds. During that time, they draw some nutrients or water into the system and then die. This is happening thousands of times every minute. What appears static—a tree standing in dirt—is actually bustling with activity. The trees are fully awake and alive. There are relationships and activities that we cannot even imagine occurring.

Under the surface we find the activities and relationships of roots, bacteria, mycelia (fungal threads), nematodes, insects, worms, and countless microorganisms. These very small creatures live in prey-and-predator relationships just as we see in the bigger animal world. In healthy soil, life is bustling. These creatures number in the millions in every teaspoon of dirt. They eat, poop, pee, kill, and die. Their effect on the soil is the green world we see around us. Without the activities of fungi and microorganisms, plant life would be very different, if it existed at all.

As growers, we create the habitat for the microorganisms and they in turn do the work of feeding the plants. If we put compost or mulch down, it is the microorganisms who make it available to the plants through complex relationships.

It is my job to feed the bustling system of soil on my farm. I do this by creating the best habitat possible. I protect the soil with a blanket of mulch or of living plants. Bare soil freezes hard and fast. It compacts when rained on. Nutrients are volatilized by the sun. They literally get sucked up into the sky. Carbon and nutrients vaporize and leach out of bare soils much faster than covered ones, maybe more than 1,000 times faster. Carbon for soil microorganisms is habitat. It is the building blocks of their ecosystems. It provides them with the infrastructure they need to thrive and live. To encourage microbial activity, protect and add carbon.

## pH and Nutrients

I do not use soil tests or measure pH anymore. Most garden books will tell you to do this, and I used to. I have found that a rich soil high in carbon content is all that is needed for plants to thrive. I add manure, mulch, leaves, hay, straw, nut shells, urine, sticks, charcoal, ashes—everything—to the soil. I am focused more on habitat for microorganisms than anything else. If you want to learn about pH and micro- and macronutrients, there are plenty of good books out there. Just remember, soil is like religion: Everyone has their own take on it and everyone thinks their way is right. I don't care about soil tests, but you can if you want to. They might really matter. I don't know everything, just what has worked for me. For now, I'm just going to add lots of carbon and keep the ground blanketed in plants or mulch.

Sometimes there are specific deficiencies for particular plants. Leaves are the best soil test in these cases. Pale leaves with darker green veins means iron chlorosis and that the pH needs to be lowered. Small, yellow leaves or stunted growth means either lack of air or lack of nitrogen. Different mulches and soil amendments will raise or lower pH and add specific nutrients. More on that in a few pages.

## Nitrogen

Nitrogen is like rocket fuel for plants. It spurs them on to vigorous growth. Too much nitrogen can lead to serious problems, though. Excess nitrogen leads to nutrient runoff, aphids, and the inability to harden off for winter. With too much nitrogen, especially late in the season, trees will keep actively growing until they are burned by frost. They really need to slow down and stop growing so that they convert new growth into wood before a freeze. A lack of nitrogen, on the other hand, will

Nursery beds at Twisted Tree Farm. A blanket of wood chip mulch promotes fungal activity and provides habitat for the microorganisms in the living soil.

leave plants stunted; some will hardly grow at all. Nitrogen needs to be balanced. Carbon is the ultimate buffer for excess nitrogen.

Because I add more sawdust and wood chips to my soils than anything else, I often hear people exclaim that I will rob the plants of available nitrogen. I have not found this to be the case. If I do ever feel that plants are lacking in nitrogen, then I simply add nitrogen. Chicken manure, fish emulsion, and urine all work fine. Be very careful, though, when applying high-nitrogen fertilizers. Organic or not, they should never come in direct contact with roots. Apply them to the top of the soil and let them leach in and be carried in by microorganisms. If roots are in direct contact, they will usually die. The best place to apply nitrogen fertilizers is just under the mulch layer. The interface between mulch and soil is the area of highest microbial and fungal activity. The mulch will also protect the nitrogen from volatilization and leaching.

## Bacterial Versus Fungal

Soils can be dominated by bacteria, fungi, or equal amounts of each. Bacterial soils exist where disturbance occurs. Plants that do well in bacterially dominated soils are generally annuals. Trees, especially species that live in older forests, form strong fungal relationships. The ratio of fungi to bacteria in an old-growth forest is heavily in favor of the fungi. In a cornfield or vegetable garden, it's the opposite.

Trees form relationships with fungi. The different types of fungi in the world can be broken down into three main categories based on how they interact with carbon: parasitic, saprophytic, and mycorrhizal. The parasitic fungi feed on living plants, the saprophytes feed on dead plants, and the mycorrhizal form a symbiotic relationship with roots, feeding them and being fed by them.

We can tip the scales of a disturbed soil toward fungal activity by adding mulches—especially wood chips, sawdust, and leaves—and by not digging anything up. Bacterial soils are not bad; they are just not as preferred by most trees. Many trees are very good at converting soils from bacterial to fungal and setting the stage for a forest to grow. These trees are the pioneer species that can compete with thick grasses and goldenrod. Here in the Northeast, white pines, red maple, black locust, and quaking aspen do the bulk of this work. They fill the soil with woody roots and cover it with leaves and sticks, transforming bacterial soils into fungal ones.

Some folks inoculate tree roots with mycorrhizae when planting. You can find the best inoculant in the soil under old trees of the same species you are planting. You can also just add a wood-based mulch and not disturb the soil after planting. Whatever you do, encourage fungal activity as much as possible whether through inoculants, lots of carbon, or both.

# Weeds

Weeds are a major consideration when you're planning out a planting. It is easy to lose a small tree in the weeds. Mowing around trees really helps get them established—it is the minimum weed control. In a lot of my plantings, I just mow next to the trees two times a year. It's amazing how much that alone can help.

Weeds are also greatly reduced by the type of soil disturbance that happens before planting. Digging up sod and flipping it over is a lot better than just planting into it. Plowing results in about 10 times fewer weeds than rototilling. Ground prep can make a big difference, but the biggest difference is made with mulches.

Young hazelnuts mulched with grass clippings. The aisles are mowed and the weeds around the trees are flattened out by the mulch. The weeds will continue to grow out and away from the tree. If the weeds were cut, they would come back faster.

# Mulch

It is impossible for me to overstate the value of good mulch. It completely changes the soil by creating good habitat for microorganisms and worms. Tunnels made by fungi, insects, and earthworms will travel deep into the earth below any layer of mulch that is not treated with dye or other poisons.

## Planting

Mulch is the key to my whole system of growing trees. It retains water, feeds the fungi, protects the topsoil, and suppresses weeds.

Many growers will talk about weed seeds in mulch, but I really don't worry about it. There's already a million weed seeds in every few pounds of soil. Mulch loosens soil so that it's easier to pull weeds.

Once large weeds are growing, I don't cut the ones right next to the tree. I have found that they grow back too fast. Instead, I flatten weeds right down with mulch. This will keep them suppressed far better than cutting or pulling and breaking roots. Applying mulch a couple of times a year for the first few years will have very noticeable benefits.

However, mulch can be heavy and expensive, which is why I love to use grass clippings from on-site.

Mulches are best applied at least to the edge of the canopy's dripline. With all mulches, it is better to make them thicker in one spot rather than spread out thinly and evenly. Thin mulch offers little benefit. I would rather have one side of a tree well fed than the whole area poorly fed. Thick piles of mulch are where moisture can truly be retained and where fungal activity can flourish.

Here's a list of different mulches. They all have different positive and negative attributes.

### Grass Clippings

This may be my favorite mulch. They are very high in nitrogen and do an amazing job at suppressing weeds. A 6-inch layer of fresh green clippings will kill just about any plant it's on top of.

In my larger plantings, I grow trees in hedgerows. I mow the alleys between the hedges two to three times a year. Going through and raking up these thick, long clippings is fast work and very satisfying. It is a practical way for me to mulch acres of trees without large inputs. Grass clippings encourage a good mix of fungi and bacteria. The best part of using grass clippings is that you don't have to haul anything in. This mulch grows right there on-site.

### Raw Manure

If raw manure is buried in the soil, it can kill plant roots very easily. If it is applied on top as a mulch, it will feed the biology very well. Most animal manures come with bedding. Applying this mix of manure and bedding before it decomposes has significant benefits. Everything that leaches out of it will be there for your plants. When people make finished compost, there is significant leaching of nutrients (especially nitrogen) during the composting process. The most nutrients available in manure are in the freshest material possible. This is how manure applications work in nature: It's spread fresh on top of the ground. Manure is a good weed suppressant, moisture retainer, and fertilizer.

Any type of manure is fine: chicken, cow, horse, sheep, and so on, can all be applied raw on top of the soil around trees. Be aware that most manures available will have antibiotics and dewormers in them. My friend Sean from Edible Acres was just telling me about a massive earthworm die-off he witnessed after applying horse manure from a non-organic source.

Manure encourages a healthy mix of bacteria and fungi.

## Compost

Finished compost is an excellent mulch, though probably the most expensive one. Compost greatly improves the texture of the soil and feeds the biology very well. It holds moisture and will temporarily block weeds. It encourages a mix of bacteria and fungi.

## Ramial Wood Chips

These are made from grinding up the tops of trees. Ramial chips are full of twigs, buds, leaves, and the like. They break down into the most incredible soil amendment over time. If applied on top, they will not rob any nitrogen. Ramial wood chips are my personal favorite mulch for what they do to the soil. They're not always available where I live, though, and are time consuming to spread compared with hay or grass clippings. Using ramial wood chips encourages primarily fungal activity, but also some bacteria and worms. It somewhat discourages tunneling rodents compared with hay or sawdust.

## Sawdust

If you live in an area with sawmills around, this is going to be one of the cheapest, most available mulches. Sawdust is light and easy to handle. Over time it develops some of the strongest fungal networks I've ever seen. I have seen piles of sawdust that were left undisturbed for a decade. It was almost impossible to break these piles apart because the fungal networks were so intertwined throughout. Sawdust holds moisture very well. It is also very conducive to tunneling rodents, who seem to love living in it.

## Bark

Shredded bark mulch is widely available from sawmills and garden centers. Most garden centers sell it dyed various colors for some reason. Bark mulch encourages some fungal activity as well as minimal bacterial activity; it breaks down very slowly. It is probably my least favorite mulch for feeding the soil, but it does suppress weeds well and retain moisture. Many bagged mulches not only are dyed but also contain chemical fertilizers, herbicides, and antifungal agents. I strongly advise steering clear of these treated mulches.

## Hay

Hay is assailed by almost every garden author as an evil mulch that will infest plantings with weed seeds. My experience has been very different. Hay is a wonderfully cheap mulch, especially if purchased as spoiled round bales that are no longer fit for animal consumption. Almost all animal operations have spoiled hay somewhere that they will sell super-cheap. Hay is very fast and easy to spread. It holds moisture in well and turns into compost quickly. The carbon-to-nitrogen ratio in hay is around 30:1—the same as you'd use for making compost. Hay does contain weed seeds, but so does all soil. Most weeds will pull out easily or can simply be mulched over with more hay. The big drawback to hay is that it encourages rodents and slugs. Hay encourages a healthy mix of fungal and bacterial life.

## Straw

A highly expensive alternative to hay. Straw is the stalks of grains, while hay is mostly grass and weeds cut from a field. Like hay, straw will also be full of seeds, though they are usually grain seeds. Some of these are difficult to pull and get rid of. I see no less weed pressure using straw versus hay. Straw will encourage rodents, but not slugs.

## Leafy Plants

The chop-and-drop system works well with large leafy plants. Comfrey is a good example of a plant that can be grown for mulch in the vicinity of tree plantings. It can be cut and left in place or collected into mulch piles. Laying down large green leafy plants will feed the soil very fast. Pile them up into larger heaps if you really want to suppress weeds. They will feed a mix of bacteria and fungi.

Aquatic plants are also a wonderful source of mulch. They usually grow fast and can be harvested several times a year. A pile of cattails laid around the base of a tree will do a great job at suppressing weeds and retaining moisture.

## Cardboard/Paper

This does a great job at blocking weeds and retaining moisture, especially if you pile another mulch on top of it. The biggest problem with cardboard is that rodents prefer to live under it more than any other mulch. You can almost guarantee a good population of voles by laying cardboard down. If you use it, be sure to pull it back away from the tree in the fall. The other downside to cardboard is that it will block rainwater when it is dry.

There are endless other mulch alternatives; your choices are limited only by your creativity and available materials. If it's convenient, I like to mulch young trees with what the older ones provide. Hickory and walnut hulls are a great mulch for those species. Chestnut burrs and leaves are good for them. Young pines get pine needles. There's no science behind this, but I believe specific types of mulch encourage specific types of microorganisms and fungi that are suited to those trees.

# Herbicide

Herbicide is the primary weed-suppressing tool in our society today. There are many chemicals that can kill plants. Agent Orange was perhaps the most famous. It was used to kill entire jungles that hid Vietcong fighters. Today Agent Orange is commercially available in a slightly altered form known as 2,4-D. It is widely used by farmers, but the most popular herbicide in the world is glyphosate, commercially known as Roundup. It is a registered antibiotic. So if you are interested in feeding the soil biology, then this has the opposite effect. Overuse of antibiotics is known to create antibiotic-resistant bacteria. Many people are aware of overuse in the medical system, and some are aware of the standard practice of adding antibiotics to animal feed, but few people realize that we are spraying millions of acres with antibiotics every year.

To me, the saddest thing about glyphosate is how effective it is. I have heard the stories of resistant weeds showing up, but around here, it works very well. Farmers' fields are clean. The only thing sprouting out of a 1,000-acre soy field that I can see are the stalks of leftover corn plants that were genetically modified for glyphosate resistance.

This lack of weeds in crop fields has been a huge change to the ecology of the world. Where once there were wildflowers creeping in around edges and in patches here and there, now there is nothing. Monarch butterflies can glide over 95 million acres of GMO corn in America's heartland and not find a single patch of milkweed.

Unfortunately, it does not stop at the farm fields. Roundup is available at your nearest box store. It is used extensively on sidewalk cracks, under guardrails, at building edges, along railroad tracks, and in all the other nooks and crannies where billions of plants used to live. As a civilization we are, for some reason, covering the Earth in antibiotics and wiping out wildflowers. If you still choose to use glyphosate after reading this, then please take some pictures of the monarch butterfly for your grandkids.

## Fertilizing

Adding mulch and compost feeds the soil, but fertilizer gives it a huge boost. It's like the difference between eating a sandwich and drinking coffee.

Fertilizing is not necessary with trees, but it does significantly spur growth. Fertilizers can really help trees become established, especially if you are trying to get them past the needs for supplemental watering and deer protection a little faster.

Organic or synthetic, all fertilizers can burn roots if they come in direct contact. For this reason, only apply fertilizer to the top of the soil. The best practice is to sprinkle fertilizer down and cover it with mulch. The mulch will protect it from washing away. Fertilizer runoff is a serious issue for waterways. It leads to tremendous algal blooms in the oceans on a large scale and is detrimental to sensitive aquatic life on a small scale. Placing mulch over fertilizer allows microorganisms to get to work incorporating the fertilizer into the soil. The area of the most activity in the soil is just under the mulch layer. Microorganisms will make fertilizers available to plants in a safe way that will not burn their roots. To me, fertilizing is more about feeding soil biology than anything else.

There are many kinds of fertilizers and soil amendments. Here's my take on a few of them.

## Synthetic Fertilizers

These are widely available at garden centers. You've probably seen Miracle-Gro if you've ever looked for anything at a garden store. Synthetic fertilizers are derived from salts and petroleum. They will give plants a very strong boost, but at a cost. Salt is not exactly the best soil amendment to apply year after year. Earthworms will leave a soil that has synthetic fertilizer applied to it. I think of synthetics as junk food. I like to eat candy sometimes, but I know it's not good for me. I'd do better to eat a dinner of meat and vegetables than a meal of gummy worms.

## Guano

This is poop from birds or bats. It's very nitrogen-rich. If you compost it first, then you will lose most of the nitrogen to leaching and volatilization. I apply raw chicken manure directly onto the ground around trees. As long as you don't bury it, roots will be safe from being burned.

## Urine

Human urine is rich in nitrogen and many other elements and minerals. Understanding this deepens my relationship to the plant world. Not only am I exchanging air with plants, but they are able to feed on my waste. I collect my urine in a jug when I'm inside and pour it around the base of trees and bushes I care about. If you apply too much in one area too many times, you will kill any plant by overfertilizing. Dogs are good at that.

## Blood

Dried blood meal is a by-product of slaughterhouses. It is an extremely nitrogen-rich fertilizer. One consideration is that blood, bones, or manure from conventional animal operations will most certainly contain antibiotics.

## Bonemeal
Another by-product of the livestock industry, bonemeal is very rich in calcium and has great benefits for the soil, especially promoting root growth.

## Soy and Alfalfa Meals
These are just ground-up soy or alfalfa. They are very affordable, though it's difficult to find the organic versions. They offer a balanced, effective fertilizer.

## Fish Emulsion
This is a liquid fertilizer. It is diluted in water and sprayed on plants or used as a soil drench. When sprayed on leaves, you can sometimes see a noticeable effect the next day. With all sprayed fertilizers, it's optimal to apply in the early morning or at the end of the day. I have heard that it's not a good idea to spray in direct, strong sunlight.

## Compost Tea
This is the best way to stretch compost. You can buy compost tea or make it yourself. Many growers will tell you it has to be brewed an exact way and studied under a microscope. I don't think it matters that much. If you have a pile of manure and a puddle next to it, then you have nutrient-rich compost tea. Use it as a soil drench or spray it directly on plants.

You can make better compost teas by brewing them. Place some raw manure or finished compost in a bucket of water and aerate it for 24 hours with a fish tank bubbler. You can add unsulfured molasses to spur microbial activity. You can also just stir the bucket with a stick a few times. Don't let the water get anaerobic or stinky, because then you are brewing pathogens instead of beneficial microbes.

## Comfrey Ooze
Comfrey is an easy enough plant to grow in the North; anyone with a few trees should grow a patch. It will do extremely well anywhere in the sun or shade in my northeastern climate. Comfrey leaves can be used

just as a chop-and-drop mulch, or you can make them into a fertilizer that is rich in calcium and nitrogen along with dozens of other minerals and nutrients.

To make comfrey into a liquid fertilizer, fill a pot or bucket with comfrey leaves. Make sure there are some holes drilled in the bottom. Place another bucket underneath. Place a stone on top of the leaves to weigh them down. Keep out of the sun. Wait a few days to a week or so. The collection bucket will contain a black liquid that you can dilute and apply to plants or the soil.

### Wood Ash

Wood ashes add an abundance of trace elements. They are high in potassium and phosphorus, but have no nitrogen. Wood ashes raise the soil's pH, and are best used on plants that prefer these conditions.

## Terra Preta: The Story of Biochar

Tropical soils are often notoriously difficult to keep fertile. They are pounded by monsoon rains that leach nutrients and intense heat that volatilizes organic matter. Modern agriculture has had a very hard time not degrading tropical soils to the point of oblivion. In recent years archaeologists in the Amazon have discovered deep black soils occurring in concentrated areas. These spots are called *terra preta*, meaning "fertile ground." In some cases the black loam extends 6 feet deep and covers several acres.[1]

Upon close examination, it was determined that these soils were human-made. They contain large amounts of pottery shards and charcoal.

It had been previously believed that civilization never developed in the Amazon because of how poor the soils are. With the discovery of terra preta, however, archaeologists have begun mapping out an enormous complex of cities that once stretched throughout the central Amazon basin. Some of these cities were the biggest on Earth at the same time that Egypt was at its climax. This civilization was wiped out by disease before European explorers saw it. Smallpox and influenza spread faster than the missionaries who first brought it. Large cities of wooden buildings disappeared into the jungle floor as fast as the people who once inhabited them. The indigenous people of the Amazon who first met the conquistadores were the survivors of a civilization that had just collapsed.

This lost civilization knew something about stabilizing soil that our civilization is just now learning about. Carbon is the building block of soil. It is what holds the nutrients and creates a soil texture or consistency that allows for simultaneous drainage and water retention. Carbon in the soil also provides crucial habitat for fungi and microorganisms. But carbon can be fleeting, especially in a place like the Amazon.

To stabilize carbon, these ancient people made charcoal, which can last for centuries or longer. They mixed charcoal with food scraps and waste to create the terra preta that still exists today.

Carbon moves through soil and mixes with air. It is drunk by plants, who build their bodies with it. When a tree trunk falls on the ground, it begins to decompose. Some carbon will feed the soil, but most of it will volatilize back into the atmosphere. Charcoal stabilizes carbon. If we are looking at the carbon cycle in regard to climate change, then biochar is one of several solutions. Here we have a way to stabilize carbon and improve soil at the same time.

Making biochar is as easy as burning carbon and putting the fire out before it turns to ash. There are many fancy methods for making biochar, but the more I learn about them, the more I am happy with my current simple cone pit. It's just a big hole—twice the size of a bathtub. I light a fire in the bottom and pile on brush. Smaller-diameter sticks work best. I pile it on and pile it on, keeping the flames very hot. The hotter the temperature at the top of the fire, the more it will rob oxygen from the bottom. After several hours of continuous active burning, the pit will be filled with charcoal. The coals are so hot that it takes a surprising amount of water to put them out. If they are not completely out, then they will continue to burn after you walk away and you may return to a pile of ash.

Most of what I burn to make char are exotic shrubs like honeysuckle and multiflora rose that I am clearing anyway. Almost everyone who manages land has some form of waste carbon that could be converted into biochar. Tree limbs, weed stalks, cornstalks, prunings, and nut shells all work. Just about any plant-based carbon source can be used in this way.

By itself, charcoal has zero nitrogen. It is a sponge and will soak up any nutrients. If you bury charcoal in the soil and grow plants on top, the plants will starve. The charcoal will suck up all excess nitrogen. It must be infused with nutrients to become biochar instead of just charcoal.

There are many ways to inoculate charcoal with nutrients. It can be mixed with compost or manure for a season, for example, or soaked in a liquid fertilizer overnight. There are no exact formulas listing how much nutrient to add to charcoal, or for how long. If plants look stunted, then they need to be fed. Sooner or later the char will get its fill and provide some of the best habitat in the soil possible.

It is worth noting that the lost civilization that developed the terra preta soils also relied heavily on tree crops. The forests surrounding these ancient cities are dominated by ancient fruit and nut trees at a much higher rate than the rest of the area. This Amazonian civilization that vanished centuries ago has left in its wake stable, fertile soils, enormous food-producing trees, and some valuable lessons for us.

## Silt Traps

Another lesson from an ancient culture comes from the remarkable farmers of China. In F. H. King's classic book *Farmers of Forty Centuries*, ingenious methods of maintaining soil fertility through intensive farming are outlined. There were many aspects of the lives of the people in this book that blew me away, but it was the silt traps that most captured my attention.

Along the courses of ditches, you can collect silt instead of allowing it to run off. Simply dig a hole at any point in a ditch or create a small dam. After heavy rains, silt will collect there, sometimes in significant amounts. Silt is wonderful soil. It is full of nutrients and it provides great texture for conditioning soils. Silt is finer than sand and coarser than clay. You can add it to garden beds or top-dress trees with it.

Between silt and biochar, you can build incredible soil out of ditches and sticks.

## Uneven Ground

Here in upstate New York there are hills everywhere, but the ground upon them is flat. The fields of our hillsides are easily mowed and grazed. They were smoothed out a long time ago when they were first cleared and plowed. It did not always look like neat rolling hills here.

Ancient hardwood forests created a ground so uneven, textured, and three-dimensional, it resembled mogul ski runs. These types of forest floors still exist today on the steepest hillsides that have been forested for centuries. Walking in an old forest around here is more like climbing up and down pits and mounds of soil than strolling down a trail. The topography is so intricate that one hillside is actually made up of thousands of micro hills and valleys.

This micro topography is created when large trees topple over. A giant root mass is lifted into the air, and as it breaks down, it turns into a mound. The place where the root ball used to be is a pit. Pits and mounds, pillows and cradles, don't sound like much, but they are. These pits and mounds have far-reaching, powerful consequences.

Before I explain exactly why I love the pits and mounds of old forests, let's take a look at a smoothed-out hillside (treed or not) The ground is easy to walk on and drive on, and it's easy to get machinery in and out, but other than this convenient access, I cannot see any other benefits. Runoff of rainwater and nutrients is constant, and can be especially severe when the ground is frozen and there is no snow cover. Streams downhill fill up faster than they can handle and flood their banks.

The soil on my farm is classified as Volusia. It is a dense clay. In the spring, winter, and fall, the soil is saturated. *Muddy* is probably the best word to describe the soil on our hill for most of the year. During a dry summer, it can be as hard as concrete. Under these anaerobic conditions, trees languish. They grow very slowly, if at all, and they usually have much shorter life spans compared with trees grown in well-drained soils.

How did a soil like this once support the massive trees that became the beams of our old house and the collapsed barn nearby? How do Volusian soils support the healthy forests all around my fields?

It is true that most of the topsoil was lost when the first European settlers here tried to plow the hillsides and grow annual grains like buckwheat and barley. When I first began planting trees here, I assumed that by adding compost, topsoil, or mulch I could improve the soil and bring patches of it back to its former greatness.

I would dig a hole, add copious amounts of compost, and apply lots of mulch on top. After a single year this spot's soil would be exactly the same as all the soil surrounding it: wet and muddy. More compost and mulch yielded the same results. What was I missing?

The answer is air. As springs bubble from deep within our hillside, the water runs through the soil and completely fills all available pores and capillaries. There is not enough oxygen in a saturated soil for roots to breathe, or for certain microbial life to flourish. In some places the soil is so wet that it smells anaerobic and has a gray-blue color. Compost might work if I added huge amounts, but sooner or later the anaerobic conditions would triumph over my soil amendments.

I realized the best way to get air into the soil during a walk in the woods. It is hard to describe the beauty of New York's southern tier. Steep hillsides covered in trees and pastures make up this idyllic landscape. It resembles hobbit country. This is the northern edge of the Appalachians. During my walk, I entered a mixed hemlock forest. The pits and mounds were so plentiful that there was almost no place that was not a pit or mound. Many of the mounds were taller than me, and some of the pits equally deep.

I have spent a lot of time looking at the ground in these types of forests. Trees are almost always growing out of the mounds. The pits fill with water and snow for large parts of the year and dry up in the summer. This system is truly remarkable: a network of raised beds and vernal pools.

The raised beds allow drainage to occur. This keeps the crown of tree roots safe from saturation by raising them above the water table. The soil in these mounds is always crumbly, wonderful stuff. And the pits are more than just water storage vessels. They slow water down, give it nowhere to go except to gently infiltrate the soil. They are not deep enough to hold water year-round. This seasonality is a big part of what makes them so exceptional.

A pond that holds water 12 months a year will inevitably become colonized with fish. Birds' legs will carry eggs from one body of water to another. A pond or pool that dries up every summer will not support fish, which can be a great thing from an ecological perspective. I love fish, raise them in my ponds, and have gone fishing my whole life. The thing about fish, however, is that they love to eat. Fish will decimate populations of amphibians and invertebrates in a small pool. The biological diversity of a vernal (seasonal) pool is very high in the

# Planting

absence of fish. These are the places where creatures like salamanders, toads, frogs, and myriad insects will lay their eggs and complete life cycles. The forest I was walking through contained thousands of vernal pools in just a few acres.

It also contained thousands of huge mounds that were growing big trees. Trees like red oak that could never grow in the wet field adjacent to this forest. The adjacent field is the same soil series as the forest: Volusia, or muddy. The difference is the uneven ground.

With this observation of forest soils, I got to work destroying my fields. They no longer have the smoothed-out surface of a hay field. I began making my own pits and mounds. After starting out with a rototiller, I have since moved on to front loaders and plows, but I also use a shovel and pick in areas too steep for machines. Without any compost or soil amendments, the trees I planted grew 10 times or 100 times faster than trees I planted into flat ground. The trees I planted in the flat ground never grew more than a couple of inches, but the mounded trees have been very vigorous; many chestnuts have been putting on 2 to 4 feet a

Typical pits and mounds in the woods created by trees uprooting and toppling over.

Chestnut growing on a homemade pit and mound.

year. Chestnuts require excellent drainage to thrive. The field they are growing in has been wet with a very high clay content. The trees often have large 12-inch-deep puddles next to their mounds. When I first began planting trees here, my neighbor declared, "The only thing you can grow here is frogs!"

With pits and mounds, we can raise frogs *and* trees.

Since building pits and mounds, I have moved on to swales and berms. These are built upon the same concept, but take it a step further. A swale is a ditch that is dug along the contour line. A normal ditch carries water downhill, but a swale catches the water and slows it down, holds it in place. The water sits and collects; most of it slowly seeps into the ground. Building a berm instead of a mound allows me to plant many more trees. I can plant long rows of trees with their root crowns elevated above the water table while they have access to water at the same time.

This is an exciting system because it allows me to grow all kinds of trees in a field that formerly could only support a few species. After just a few years, tree roots will extend beyond the mounds or berms. They will find the surrounding ground. This was a big concern of mine when I first started out. So far, I haven't seen the trees slow down

# Planting

*Left,* A couple of swales built along the contour lines of the hill. Trees are planted along the top berms and water is collected and evenly distributed in the swales. *Right,* Young chestnut seedlings planted on a berm.

at this point; rather, they are speeding up as their root systems build strength. I believe this is due to two factors: Their root crowns are the most important part needing drainage, and the surrounding ground is not the same. The water table at the surface has been altered by numerous pits and swales. Water no longer runs unimpeded through the top foot of the soil. There are traps and sinks for it everywhere. The ground below the swales is not as wet as before.

I'm sure that there are many consequences of these types of earthworks that I am unaware of. Using the forest as my example, I believe that uneven ground fosters growth and biodiversity on many levels. I hope that numerous fungi, bacteria, insects, and amphibians are finding the little niche pockets that they need to thrive.

Forest floors were created through an unbelievably complex and rich system. For those of us trying to grow trees in old fields, uneven ground can similarly give our trees the boost they need. Automatic mulching in the fall, raised beds, and water storage are just a few benefits of such a system.

# Interplanting

Mixing in a diversity of species adds to the resiliency of any planting on multiple levels. Not every species will have a good crop every year. However, with a multitude of tree types planted, there will be something that is having a good crop just about every year. Mixing species will not only ensure that something works out well, but also confuse pests and increase biodiversity.

Looking to nature as an example, we see mixes of tall trees standing over shrubs, standing over herbs, and standing over fungi, with vines mixed throughout. Imagine a collection of tall nut trees with fruit trees around the edges, and shade-tolerant berry bushes like currants and raspberries forming the understory over a bed of medicinal herbs and edible mushrooms.

Creating a system like this takes a deep understanding of plant habits. It's easy to have a lot of plants close together competing for light and not producing a lot of food. It takes nuance and art to keep all these pieces working together symbiotically.

Once your plant identification skills are sufficient, you can begin to see which plants support each other in nature and then use that knowledge to create your own diverse plantings.

Growing crops underneath the canopy of larger trees is widely used in the tropics on a commercial scale. Coffee, chocolate, taro, and many other plants do well under gigantic tropical fruit and nut trees. It is a beautiful system to see trees providing the protection and shade the smaller plants need, while the smaller plants blanket the ground, creating a living mulch.

## Berries in the Shade

Simply interplanting trees and shrubs does not mean that there will be extra food to harvest. Only some shrubs will flower and fruit in the shade. Almost all of them will fruit less in the shade than in the sun, but some will still produce nice crops. The most shade-tolerant, productive fruiting shrubs I've worked with are cornelian cherry, currants, gooseberries, bunchberries, hobblebush, and raspberries. There are many other possibilities, but these species are particularly well suited to fruiting under other trees.

# Planting

## Polyculture at Edible Acres Nursery

Sean Dembrosky of Edible Acres is a wizard with interplanting. He has developed a deep understanding of the relationships among different plants through thousands of hours of observation and work in his food forest and permaculture nursery. When Sean plants a tree or shrub, a whole ecosystem is planted.

Here, in this 10-by-12-foot elderberry planting, Sean has rhubarb acting as a rhizome and deer block. The spring-flowering bulbs that are planted under the rhubarb pop up in early spring before the rhubarb and die back when it is fully leafed out. Echinacea and marsh mallow are planted throughout for pollinator support and diversity. Garlic is mixed in for pest confusion. Three rows of currants to the left of the large elderberry include red, black, and white currants. On the southern edge (which cannot be seen in photo) are wild bee balm (*Monarda fistulosa*), sea kale, and sage. Wherever there is any room, ramps (wild leeks) are planted underneath. Now, that is interplanting!

Sean tells me there is no end. He can always see where another layer can be woven into a planting.

Photo courtesy of Sean Dembrosky.

Mulberry is one of the only fruit trees that can produce in the shade. I've seen mulberries laden with fruit under a closed canopy of black walnuts.

It's important to note that not all trees produce an equal density of shade. It takes observation to learn which trees provide heavier or lighter shade. Walking through the woods, you can see where the ground is just leaf litter and mushrooms, and where it is a thicket of shrubs and seedlings.

White pine, black and honey locust, cherry, and ash are just a few examples of trees that do not create a dense canopy. Maples, spruce, hemlock, and beech are some of the species that will block most of the light. Mushroom cultivation and medicinal herbs are better suited to living under some of these denser-shading trees.

## Nitrogen Fixers

Another component of interplanting is using nitrogen-fixing species that will benefit the rest of the planting. In the shade, the nitrogen gathered by these plants will be greatly reduced. However, if they are allowed sufficient sunlight, the benefits to the soil and the surrounding plants can be significant.

Nitrogen fixers work by forming associations with specific bacteria that can harvest nitrogen out of the atmosphere. You can actually see this nitrogen most of the time if you dig up plants. There will be small round nodules along the roots that are filled with nitrogen.

Nitrogen fixers don't always share nitrogen with surrounding plants. They generally only do so when some of their roots die and are sloughed off into the soil. You can create these root die-back events by mowing, grazing, or cutting branches during the growing season.

The most famous nitrogen fixer is clover. It is taprooted and highly adaptable. There are many types of clover ranging in height from a few inches to a few feet. They have deep-penetrating taproots and form bacterial associations easily. Clover seed is cheap and easy to establish. You can purchase it and add inoculant—the bacteria that fixes the nitrogen. However, I have found that these bacteria are abundant enough in the soil that associations are formed within the first season without inoculant most of the time.

There are several other herbaceous nitrogen fixers, including bird's-foot trefoil, vetches, peas, and beans.

Some trees and shrubs can also fix nitrogen. Black locust is the biggest, most robust species in this category. It spreads vigorously through root runners. As these are mowed off or grazed in the summer, nitrogen is deposited in the soil at noticeable rates. Siberian pea shrub, sea buckthorn, alder, and autumn olive are all shrubs that fix significant amounts of nitrogen. All of these species require abundant sunlight to do their work.

## Taprooted Biennials

These are herbaceous plants that have a life span of two, sometimes three years. In their first year they make a basal rosette of leaves and a deep fleshy taproot. The second year they send up a stalk and flower and go to seed. Once they have matured their seeds, they die. The roots that punched so deep into the soil and subsoil are left to rot. I think of taprooted biennials as a way to deposit carbon where I cannot reach. They spike the ground, aerating the soil and improving it deep down.

Some of these plants also make lots of flowers for pollinators. Members of the carrot family, the clovers, dandelions, and chicory are some of the best-flowering taprooted biennials. The most powerful roots come from the docks and teasel. Burdock is supreme in its ability to penetrate deep into the subsoil. All the species of *Rumex* (yellow dock, curly dock, et cetera) are also incredible in their ability to penetrate clay hardpans.

All of these plants do just fine in the most compacted, heavy clay soils, or in dry sandy soils. In both cases they will add organic matter and aerate the soil. They are as easy to grow as sprinkling seeds around bare earth.

## Pollinator Plants

Certain perennials produce copious nectar flows throughout the summer. Some of these will flower almost the entire season. Many of them have very tiny flowers that are loved by beneficial predatory insects. There are countless wildflowers that support pollinators and beneficials. Members of the carrot family are some of the best. The mint family impresses me every year with the level of bee activity. I think growing mint around the base of fruit trees and young nut trees is one of the most pleasant companion plantings. There are hundreds of varieties of mint, and there is no reason not to grow as many as possible. They are care-free, they're not eaten by deer, and they produce some of the most popular nectar for bees.

## Spring Ephemerals

These are perennial plants that gather the majority of their sunlight during the first half of spring before the trees leaf out. Most of them thrive in the deepest, oldest forests. Many of these plants are extraordinary wildflowers and powerful medicinals. Some examples include trilliums, Dutchman's breeches, wild ginger, hepatica, black and blue cohosh, wild leeks, goldenseal, and ginseng. Most spring ephemerals are heavily browsed by deer and crowded out by exotic plants such as periwinkle and garlic mustard. Cultivating these spring ephemerals adds a valuable layer to any forest planting and is a truly positive contribution to the world. They also can provide a good stream of income, whether they are sold for their medicinal parts or as nursery stock.

## Shade Perennials

Many, many perennials thrive in full shade all season. There are several to choose from that offer medicine, food, and other benefits. Solomon's seal, false Solomon's seal, comfrey, meadowsweet, good King Henry, ferns, and lots of other species will add beauty and utility to forests and tree plantings.

# The Benefits of Bareroot

The most convenient things are not always the best things. This is the case when it comes to quality trees for planting. The differences between bareroot and potted trees are numerous and very significant on several levels. With a potted tree, we have the convenience of buying any time of year and delaying planting for months.

Bareroot trees are grown in the ground, never in any kind of container. They can only be safely transplanted when they are dormant and the ground isn't frozen. Here in upstate New York, that is a short window that generally only happens during the months of November and April. The rest of the time, the trees can't be moved without risking heavy losses. So what's so great about bareroot trees that I would be willing to sacrifice this convenience?

# Planting

## Soil

Because bareroot trees are grown in the ground, they require zero potting soil. There is no need to truck around perlite, peat moss, shredded bark, or any other bulk materials. The soil used to grow bareroot trees stays in one place forever; it never needs to be hauled anywhere. It can be built on and improved year after year. After only the last 10 growing seasons, I am deeply satisfied with my nursery beds, which are teeming with earthworms, mycelia, and a healthy network of living microorganisms. It would have been a shame to sell all that soil off in pots.

Bareroot trees have naturally shaped root systems. This is a one-year-old American persimmon that was grown in living soil.

I grow my bareroot trees in deep beds of loose, living, friable soils that are extremely high in organic matter. The soil is so deep in my beds that I can often pull out whole roots of dandelions and sometimes even burdock without a tool.

At my nursery, a layer of mulch, usually in the form of wood chips, covers the beds. This protects the soils from the compacting effects of rain, and from the volatilizing effects of the atmosphere. As rains and melting snows seep into these nursery beds, the water is slowly absorbed through the capillaries of the soil. Compare this with a potted tree, which leaches out rainwater, carrying with it nitrates and phosphates from fertilizers that can enter waterways.

I should mention that not all bareroot nurseries have the best soil practices; many of the larger ones in particular use herbicides and mechanical cultivation on their fumigated, synthetically fertilized soils. So if it is important to you, check in with individual growers. There are several nurseries besides mine doing great things with their trees and soils.

### Roots

A living network of microorganisms, earthworms, and fungal mycelia interacts with the tree. In this environment roots form fibrous, spreading systems. They do not circle around each other as in a pot, but rather form natural, healthy, and quite beautiful shapes. A circling root system can often lead to permanent damage, as tree roots will wrap around their own trunk, eventually girdling the tree. The roots of field-grown trees come in the shape determined by the tree rather than the container.

Quality-grown bareroot trees are feeding in a living soil that does not suffer extremes in temperature fluctuation. Their roots will become extremely fibrous with many hairs, and their rates of growth can often be double those of potted trees grown right next to them.

### Water

Pots need to be watered frequently. They are usually aboveground and black. The temperature in containers fluctuates constantly. A small black pot can dry out in a matter of hours. Compare this with mulched soil, which can take weeks if not months.

Almost every day that it doesn't rain during the summer, I water my potted plants. It is a very different story in the bareroot beds. Because of the high organic matter content and a layer of mulch, I almost never water them. The soils in the bareroot beds can be kept so healthy that there is no need for any irrigation whatsoever. This is a tremendous savings in energy, time, and materials. Also, these beds can absorb water, rather than leaching out water and nutrients.

One added bonus to the water aspect of bareroot trees is that they are planted while dormant. The tree suffers very little, if any, transplant shock. The soil is moist during the dormant season, and a thick layer of wood chips will keep that moisture all year during a typical Northeast summer. So bareroot trees rarely even need to be watered after planting in my climate.

### Space

Bareroot trees can be grown on very tight spacings. I grow 4-foot apple trees only about 5 to 6 inches apart within the row. I grow mulberry seedlings by the hundreds in a 3-by-20-foot strip. In a 40-by-50-foot space, I can fit well over 1,000 trees. I grow about 20,000 trees per year in my half-acre nursery. There is no way this would be possible with potted plants. By their very nature, they take up more space, especially the popular round pots that facilitate circling root systems.

Typically potted plants sit on top of plastic weed fabric, herbicide-cleared gravel, or pavement. How much of this space could be devoted to something more beautiful if we grew our trees in nursery beds instead of nursery pots?

## Price

As if the health of the tree and the soil wasn't enough to make bareroot the best choice, in addition the price is also much lower. When we buy potted trees, we are paying for someone to purchase or make potting soil, to move around a pot and overwinter it, and to water that pot incessantly. We are paying more money for a disfigured root system that may never recover.

Because bareroot trees are so much easier to grow, and so much less resource-intensive, they are a lot less expensive. Container trees cost typically two to three times more money than a similar-sized tree grown in the ground.

## Timing

This is where bareroot really suffers, or excels, depending on your perspective. Bareroot plants look like a stick with roots during a gray time of year. Potted trees can be full of leaves and flowers during the most exciting times of spring and summer for gardeners.

Bareroots are only available during April and November here in New York. By the time May rolls around, trees are beginning to leaf out, and it is not as safe to transplant them without soil attached to their roots. When we do transplant bareroot trees, it must be done carefully so that the roots can never dry out. They will come packed in a moist material like sawdust or newspaper and should be planted or heeled in immediately.

Planting trees within the timing of when they are dormant is not as convenient, but it is better for the trees. They experience very little transplant shock, and when spring arrives in full force, they hit the ground running.

A hundred years ago no one sold trees in pots, but people still exchanged and transplanted lots of fruit, nut, and ornamental trees.

Bareroot trees are much lighter than pots. It is no big deal to carry 100 of them in an armload or mail them in a box.

They even shipped these plants around the world. This was all done during the right time of the year for the trees—while they were sleeping. I encourage gardeners to return to a simpler method of obtaining trees, one that does not require plastic pots and potting soils being trucked around, and one that leaves our soils and waterways healthy. In no way do I mean to berate the growers of container trees. My intent in writing this is to illuminate the profound benefits of quality bareroot trees.

Planting trees is an amazing experience. I feel a sense of relief as they go in the ground and tremendous satisfaction looking back at my work at the end of the day. When trees are in the ground, I have something to check on, to care for. I love checking back on trees that I have planted. Sometimes there is a fence to fix or weeds to smother, but if the tree is growing, then who cares? Seeing them get bigger only invigorates me to help them more. It's really a nice feeling to see birds land in a tree that you planted, to sit in its shade, or collect fruits and nuts from it with friends and family. You have done something good for the world when you plant and establish a tree.

There will be many times when a planting does not work. That happens to every single grower. The best growers are the ones who respond to failure by making adjustments and planting more. When a planting doesn't work, there is always a reason. You may not figure it out the first time, but you probably will by the third or fourth. Just keep planting and sooner or later, you'll be so glad you did.

CHAPTER THREE

# Propagation from Cuttings

All propagation can be broken down into two categories: sexual and cloned. In sexual propagation, flowers are pollinated and seeds are planted. The result is a plant that is not exactly like its parent, but often similar. This expands the genepool of a species every time it happens. In cloning, a copy of a plant is made. This preserves every genetic trait that the parent plant has. If it has some amazing unique fruit, then it still will as a new plant. Cloning does not expand a genepool, but it does occur frequently in nature. We can find massive clones in a stand of cattails or a grove of aspens. Cloning and growing from seed both create exciting possibilities for any grower.

## Cuttings

The experience of rooting cuttings continues to amaze me year after year. It is one of the most satisfying propagation techniques. Cutting propagation can be rapid. A single tree or bush can be turned into many large plants within a single season. It is a technique that builds upon itself readily. I have taken many cuttings from cuttings that just previously rooted. The numbers can multiply pretty fast.

There are two general types of cuttings, hardwood and softwood. Hardwood cuttings are taken when stems are dormant. Softwood cuttings are taken from actively growing wood.

You want to plant cuttings deeply so that they are less likely to dry out. You also do not want to have too much top growth before rooting—in fact, the less the better. Top growth pulls water from the stem. *Little tops and big bottoms* describes how I try to keep cuttings before they root.

All types of cuttings root best when they callus. It is the same tissue that forms on a tree when you cut a branch off. This tissue is a sign of good healing from a wound. The cuttings you take also need to heal their wounds, especially on the bottom. Callous tissue is very sensitive. It is the most likely place from which roots will emerge. With some easy-to-root species, it won't matter whether they callus or not. However, callusing is essential for difficult-to-root varieties.

## Hardwood Cuttings

Hardwood cuttings can be taken anytime from fall through early spring while plants are dormant. There are nuances to different species, but the general idea is to take cuttings from vigorous one-year-old shoots. Anywhere from 3 inches to 9 feet is a possibility, depending upon species. Most hardwood cuttings I take are around 10 inches long. I plant them to their full depth, leaving only one or two buds aboveground. I plant them whenever the ground is not frozen during the dormant season. I've planted lots of them in November, during midwinter thaws, and in early spring. I have not noticed a big difference with the timing. If you have good, strong shoots from a species that roots well, you will likely have success.

Elderberries, currants, goji berries, grapes, cottonwoods, and willows are some of the easiest types to root. I often have close to 100 percent success by just sticking these in the ground anytime they are dormant. Many other species can root well, but often require a bit more help.

Before planting hardwood cuttings, it is a good idea to give them a big drink. I place cuttings in a bucket or jar of water before planting. You just need the bottom end in water. They can stay there for up to a day. If you leave them longer, be sure to change the water so that it doesn't get anaerobic. This will kill any cuttings if they are left in unchanged water for too long. Some growers will leave cuttings in water for weeks, but they will frequently change the water or aerate with a bubbler.

Hardwood cuttings are easy to handle and store. They don't dry out very fast the way softwood cuttings do. After you collect them, you can plant them right away or store them for later use. I collect hardwood cuttings over the course of the winter. Often the ground is frozen or buried in snow. I store cuttings several ways.

# Propagation from Cuttings

**Storing Hardwood Cuttings**

- They can be stored in the fridge. Wrap the stems in barely damp paper towels or newspaper and place them in a plastic bag in the coldest part of the fridge (usually one of the lower drawers). Check for mold every couple of weeks. Don't let mold get out of hand; you do need to watch for it if you store anything in the fridge. Some growers warn against putting cuttings in plastic bags because of mold. Instead they wrap cuttings in a damp cloth. I have found that they can still mold this way, and they also dry out much faster.
- Storing cuttings outside is easiest if everything is not frozen solid. I have never had any mold problems storing outdoors regardless of the medium they were in. I have buried cuttings in snowbanks, mulch piles, sand, and dirt. Anything except unfinished compost works well. You don't have to bury the whole cutting, but you can. You just need to make sure the bottom half is buried. That way the stem can still drink.
- Storing cuttings in the basement or an attached garage works very well if it's unheated. You can pack them in large pots filled with barely damp, well-drained medium. Sand is the best, but also the heaviest. Sawdust works well, but be careful: Using fresh sawdust can have adverse effects. Also, sometimes sawdust is actively decomposing while the cuttings are stored in it. It can heat up and kill entire batches. I have lost many plants because of sawdust getting too hot. If you use aged sawdust, it will work great.

You can also place cuttings on a concrete or dirt floor and cover with burlap sacks. Keep the sacks wet. Be cautious about storing plant material in basements or garages. It can lead to mold problems in a house.

Hardwood cuttings of elderberry collected in winter.

Elder cuttings processed into lengths for planting. They are planted deep enough that just the top bud is showing an inch or two aboveground.

Elder cuttings growing over the summer in well-fed nursery beds.

## Transplanting Hardwood Cuttings

Don't rush this step, I have learned. It's better to wait until roots are well formed before transplanting. It is best to leave cuttings where they are until they go dormant. Sometimes you will want to move them sooner for whatever reason. If you must transplant actively growing cuttings, then pinch their leaves back and soak them before gently moving. With elderberry I pinch all the leaves off; otherwise they often wilt and die.

If you are lucky enough to root cuttings before they have leafed out, then they transplant well. I have done this with black cap raspberries that I rooted with bottom heat in February. I planted them out in April before they leafed out. If I could do that with everything I would, but most species will leaf out before many roots have formed.

## Root Cuttings

Root cuttings are when a fragment of root is planted. You can plant the cutting horizontally or vertically. If you plant it vertically, then many growers advise planting the side that was nearest the base of the mother

## Propagation from Cuttings

plant up and the part growing away from the mother, down. I haven't figured out which orientation is best. I have had extremely variable success with root cuttings. Usually one of two things happens. First, the root fragment explodes with growth and turns into a massive plant in one year. I've had black locusts reach 9 feet in a year from a 1-inch fragment of root, elderberries that reach 5 feet with multiple stems and even flowers and fruit in one year, and raspberries that turn into a small patch with ripe berries within a few months. Or second, root cuttings can disappoint. What often happens to me is that they callus over, grow more roots, but don't ever send up a shoot above the ground. I've seen elderberries sit like that for two years before disappearing into the soil.

One-year-old rooted elderberry ready for transplanting.

Plant root cuttings close to the surface. Many species will naturally send up root suckers when their roots are close to light. Some growers dip root cuttings in rooting hormone to induce callusing. I sometimes heat the soil for root cuttings by placing them in trays on bottom heat or in a greenhouse. It's easy to forget about root cuttings after planting them, but if you pay attention to them, keep them watered, and make sure they're not lost under the shade of weeds, you are more likely to have success with them.

### Softwood Cuttings

Softwood cuttings require more care and nuance than hardwood cuttings, but they are more likely to be successful with hard-to-root varieties. You can also produce huge amounts of plants in a short time with softwood cuttings. They can root in as little as a week in some cases. Rooted plants can then produce more material for additional cuttings within the same growing season. Softwood cuttings have allowed me to root varieties with which I never thought I'd succeed.

The timing for softwood cuttings is while they are actively growing. You want to select vigorous growth. It should be firm enough that it snaps when you bend it. If it just bends and folds, but doesn't crack, then the growth is too young. Check back with it in a week or two. Most softwood cuttings are ready sometime in June in my area. It varies with species.

Taking softwood cuttings in the morning is preferable to afternoon. Cuttings will be more turgid and full of water in the morning.

With full green leaves attached, softwood cuttings can wilt very fast. If they are wilting, they are severely weakened and will probably not bounce back. I place softwood cuttings in a bucket of water as I'm collecting. I then find a place in the shade to trim their leaves back. I either cut leaves in half or cut the whole leaf off.

You can make softwood cuttings as small as from one leaf node to the next, or they can be a foot long. They are stronger when they're larger, but you get less plant material out of them. With many species, a very small cutting works fine. I've rooted haskaps and goji berries that were around an inch long.

The key to softwood cuttings is that you don't let them dry out or become waterlogged. The stems are more prone to rot than hardwood cuttings. I plant softwood cuttings in pure sand. I also keep them in the shade or under shade cloth to prevent them from drying out. They need to be watered a lot so that they don't become stressed trying to support leaves. A mist system is the ideal setup for softwood cuttings. I have only used a mist system for the last two growing seasons, so my knowledge is somewhat limited. Here is what I have learned so far.

## Mist System

With this system, a mist is sprayed on the cuttings on a timed schedule to keep them from drying out. You can purchase specific mist timers through nursery/greenhouse suppliers. They come in many forms and styles. The easiest ones to set up run on a battery instead of being plugged

## Propagation from Cuttings

in. You can adjust the settings so that cuttings are sprayed with a fine mist every 10 minutes for 10 seconds, and the system shuts off at night. You will need to adjust the settings depending on weather. During rainy, cloudy days they will need less mist. On hot, dry days, they'll need more. The key is to watch the cuttings. You don't want them to stay soaked. What you do want is for them to have a light film of water on their leaves that dries up shortly before the next spray of mist starts.

Mist nozzles come in many different options. The more expensive brass nozzles spray a very fine mist. Some of the cheaper plastic ones provide more of a leaky-hose-type spray. All of the nozzles clog sooner or later, but the finer sprays tend to clog more often. This is tough because the fine spray is the most beneficial and gentlest form of watering.

Setting up a mist system with bottom heat. Blue foam insulation lies under a snaking heat cable. Strung along the hoops is a black water line with mist nozzles plugged into it every foot. The hoops will be covered with burlap and the bed will be filled with sand.

The mist nozzles are attached to a water line, either PVC or some type of irrigation line. Attach a drain valve to the end of either type of line; otherwise the whole system will drip after each mist cycle. Dripping will cause that spot to be totally waterlogged.

You can house the entire mist system under a plastic tent or under a burlap / shade cloth tent. A friend of mine keeps his system just under the open sky on the north side of a barn. You don't want the setup in full sun without shade cloth.

I place bottom heat under my mist bed. Even though it's summer when I use it, the nights can be very cool in my area. The bottom heat provides an even warm temperature to encourage faster callusing.

There are endless variations on mist systems as creative as the growers who use them. Just keep in mind the goal: cuttings that don't dry out and aren't saturated.

Transplanting cuttings out of a mist system can be tricky. The plants are small and delicate. If you pull them out of mist and just place them in a bed, they will often die. They need a slight hardening-off period. It is a good idea to move them just to the edge of the mist for a day or two before moving them further. Place them under deep shade at first. When I plant rooted cuttings from the mist house, they go under a tent of burlap. I water them a couple of times a day until I am confident they have transplanted well. Wait at least a few weeks before removing the shade cloth. If it is later in the season when the cuttings have rooted, you may want to not plant them in a bed, but bring them inside to a protected space for the winter. For small sensitive plants, I do not allow them to freeze. They spend the winter in my basement and get planted out in the spring.

Mist systems can take a lot of tweaking to get just right, but once you have a good setup, they are amazing. I believe a good mist system and some nice stock plants could provide anyone with a very decent income and thousands of plants. It might sound like a lot of work and setup, but if you are serious about propagation, a mist system is a game changer.

## Rooting Hormone

Rooting hormones are naturally found in plants. In the old days growers used willow water. They would sometimes also insert a grain seed in the bottom of a cutting. The seed releases hormones when it sprouts that help the cutting along.

Today rooting hormone is found as a synthetically derived chemical that causes plants to form callous tissue. It is the callous tissue from which roots are most likely to form. Rooting hormone comes in powder and liquid forms and in varying strengths. Use the strength that is indicated on the box. For hardwood cuttings you usually want the strongest form. Read the instructions and be careful when handling. Synthetic rooting hormones are toxic chemicals.

I used to use rooting hormone all the time until I learned of its toxicity and links to disrupting people's endocrine systems. It was very hard for me to stop using synthetic rooting hormones. Propagating plants provides my family's income. I was worried that if I stopped using them, I would have fewer plants to sell. Before I stopped, I carried out dozens of trials, planting half a batch of cuttings with the hormone and half without. I have been gratefully surprised to learn that it almost always made no difference.

I have read about natural rooting hormones like honey and willow water. I have tried these on occasion, but have not noticed a significant benefit personally to justify the time involved. It's a lot faster to plant cuttings if you don't have to dip them first. The time adds up when you are planting thousands of cuttings. So, for what it's worth, I maintain a profitable nursery business that roots thousands of cuttings every year without any rooting hormone.

## Bottom Heat

Bottom heat can make a big difference in my experience. I use it for some seeds as well as cuttings. *Warm bottoms and cool tops* is the ultimate combination when trying to root difficult cuttings in early spring. Cool tops will keep the plants dormant, while warm bottoms will encourage callusing. If the air temperature is around freezing, cuttings won't make any top growth. But if the soil temperature is around 75°F (24°C), then they can begin to form callous tissue. Bottom heat does a great job of encouraging root activity.

In the old days nurseries would create bottom heat by planting cuttings upside down in the fall. The bottom of the cutting would be near the soil line. In the spring, when the soil warmed up, it would heat the bottoms before the tops.

Some growers create bottom heat with piles of hot manure. This is pretty neat, but it's not easy to regulate the temperature. I like to keep cuttings at 75 to 80°F (24–27°C) for months at a time.

Today it is easy to create bottom heat with a small amount of electricity. You can go to most garden centers and buy small heat mats. You can buy bigger ones online. The cheapest option is to use heat cables. Instead of a mat, the cables snake back and forth warming a bed of sand. You can also use Christmas rope lighting instead of the heat cables; just be sure it's not LED lights, as those emit no heat.

The main reason for providing bottom heat or rooting hormones is to stimulate the growth of callous tissue at the base of cuttings.

Insulate very well under the bottom of the heat mat or cable. You can cover the top of the bed or trays with a sheet of plastic and you'll create a super-humid jungle. But remember, you want to keep the tops cool and dormant.

Make sure to use bottom heat along with a thermostat. You can easily over- or underheat a bed. Anyplace that sells heat mats will also sell the automatic thermostats that go with them. These will keep soil temperature to within a single degree for months on end without you having to keep track of anything. They are easily worth the $20 they cost.

## Medium/Soil

I think of cuttings as a race between rooting and rotting. Either roots are going to form or the bark will start to allow fungus in. They can last a lot longer in a very well-drained medium. Sand is the best. Pure 100 percent sand is my favorite medium for starting cuttings that I will be transplanting soon after they root. If they are cuttings that I'm sure will root and are started in just beds outside, then I will plant them in rich soil amended with sand.

You can also use perlite, but it's not as nice as sand in my experience. Play sand or construction sand both work fine.

Don't be shy at all when you propagate, or follow any rules. You're working with living organisms. They will constantly surprise you and break all the rules that anyone will ever tell you. There are no people with green thumbs, just people who plant lots of seeds and cuttings. The more you plant, the more you will learn and the more you'll grow. Once you understand the basic principles, you can really begin to experiment.

CHAPTER FOUR

# Propagation from Seed

Starting trees from seed is in my opinion the most satisfying propagation experience. With a few movements of the hand, thousands of seeds are cast out and raked into the earth. It's possible to start 10,000 trees in just a few minutes when using seeds of certain varieties. Each tree that grows is genetically unique, furthering the diversity and strength of the species.

Growing trees from seed not only is the fastest propagation method, but can also be highly profitable. In a small space you can raise a lot of trees to sell. I have had 4-by-20-foot beds of seedlings that were worth several thousand dollars after a single growing season. Nothing I can think of (that is legal) even comes close to that kind of farming.

Starting trees from seed is not always easy, though. Many things can go wrong both before and after sprouting. There are many details I've figured out along my journey of propagating trees. I hope that reading about my mistakes over the years can help you avoid them.

## Stratification

Many types of tree seeds require a period of stratification. This big word just means that the seeds need to be cold and moist for a few months. In nature seeds go through a stratification period over the winter. If they

First day aboveground for this pear tree.

don't feel this cold period, they won't sprout. This is protection against sprouting at the wrong time of year.

There are several ways to replicate winter's cold for your seeds. The fridge is simple, but there are a few things to be aware of. Seeds stored in the fridge are prone to mold, so check on them every few weeks or so. Also, refrigerators are not as cold as outdoor winter temperatures, so sometimes seeds will sprout too early, when the ground is still frozen outside. Store seeds in the back of the bottom drawers of the fridge—the coldest spot.

I store a lot of seeds packed in tote bins in my unheated basement. Again, mold and early sprouting are serious concerns. I often open a vent on cold nights to allow cold air to rush into the basement. I close the vent during warm days.

The best place to store many seeds is outside. However, rodents are a big danger outdoors, especially with nuts and stone fruits. One way to protect against rodents is to drill holes in the bottom of a bucket. Bury the bucket and snap the lid on to keep mice out. I keep the top of the bucket right around the soil line and cover it with a thick mulch.

You can also create a mouseproof enclosure by trenching in hardware cloth. It is amazing how deep mice will tunnel, so either bury it a minimum of 2 feet deep or have the bottom of the pit lined with hardware cloth. Also, the walls of the seed enclosure should be high enough that mice and chipmunks cannot walk over when the snow is deep. I have found that 3-foot-high walls are mostly effective at keeping rodents out, especially if I trample down snow around them. If you have squirrels, you will need a top to the enclosure, which makes burying buckets seem like less work.

With any stratification method the seeds need to be packed in a growing medium to keep them moist. I think play sand is the best for this, especially white play sand because it's easy to see the seeds. Sand is the least likely of any medium to grow mold. You don't need to worry about mold so much if you are using the enclosure method. Dirt works fine if it's not in a bag or a bucket. Whichever medium you use, mix in the seeds at about a 1:1 ratio by volume of seeds to medium.

I have had seeds of numerous species freeze solid outside in my rodent enclosures, even when I mulched them. This has never stopped anything from sprouting for me, though I have heard several other growers warn against letting seeds freeze. I think if seeds are only buried under a thin mulch of leaves over the winter in nature, then we can do the same.

## Propagation from Seed

These aronia seedlings were planted in a tray and left mulched outside for the winter—the stratification process.

## Scarification

Scarification is a procedure that's used to weaken the seed coat. Some species have a hard coating around the seed to protect it. This coating can keep certain species dormant for decades or longer. Trees use this strategy in order to build seed banks in the soil that sprout when conditions change. Many trees will only grow in the sun, so they sometimes build these seed banks against the day a big disturbance comes along and opens up the canopy.

There are a few ways to weaken the seed coat. One way is to abrade it with a file. Abrade deep enough that you see a color change. Usually seed coats are black or brown. File deep enough to get to the white inside. This works if you are only planting a handful. If you're planting more, then hot water is the way to go. Bring a pot of water to a boil. Turn off the heat so the temperature is below boiling. Drop in the seeds and let them soak overnight. In the morning the seeds should look swollen and plumped up.

Some growers use gibberellic acid to weaken seed coats and germination inhibitors. Gibberellic acid is a hormone found naturally in plants, but synthetically derived and sold as a commercial product.

## Timing

Even if a tree species is very hardy, that does not mean that its freshly sprouted seed is. Many types of trees will die if they encounter frost when they are just emerging from the soil. If a frost is coming, I cover seedlings with row cover, bedsheets, or fluffed-up hay. Some species sprout as soon as the soil warms up and need to be planted right away. Others are stored dry and you can wait until after the danger of frost is past.

Apples, pears, and plums can handle some frost when they are just emerging. Chestnuts, hazels, and walnuts will be severely damaged but resprout. Locust, hackberry, and mulberry will die completely.

## Planting Depth

Generally seeds are planted as deep as their thickness. So a 1-inch-thick seed is planted an inch below the soil line. Tiny seeds can just be lightly raked into the surface of the soil. When seeds are planted near the surface, it's important that they not be allowed to dry out. Mulching seedbeds greatly protects them from this. Of course, the depth of the mulch will determine how deep the seed should be planted.

## Air-Pruned Beds

Transplanting trees is a highly unnatural act. It is hard to imagine a scenario in which a tree would be transplanted in the wild. Trees don't have any evolutionary history of being transplanted, and some species have a very difficult time with it. Air pruning is one method that I have found eases the act of transplanting young taprooted trees.

Certain species grow a taproot when their seed first sprouts. This taproot is shaped like a carrot and usually dives straight down. In some cases, taproots can reach over 2 feet the first season. Damaging the taproot when digging up a tree often results in serious injury or death. Some growers use pots to raise taprooted species, but this results in a circular root system that is not in the best long-term interests of the tree. However, there are pots on the market designed to deal with taprooted trees; they operate on the concept of air pruning. These pots are expensive and made out of plastic. I have found air-pruned beds to be a more efficient and cost-effective way of raising taprooted species.

## What Is an Air-Pruned Bed?

An air-pruned bed is simply a bed that is raised off the ground with a heavy-duty screen on the bottom. As the taproot reaches down into the soil, it will eventually find the bottom and hit the air below. When this happens, the tip of the root is pruned off very gently by drying out.

The plant's response to being air-pruned is fascinating to me. If you cut the top of a tree off, it will respond by branching out from below. If you pinch any shoot of a plant, it will make new shoots from the sides as lateral buds are awakened. There is a hormone in the tip of a branch and in the terminal shoot of a tree that suppresses buds and branches below. When that tip or leader is removed, the lower buds are released. The same thing happens underground. When the tip of the taproot is pruned off, the rest of the root responds by branching out. This action transforms a taproot (shaped like a carrot) into a fibrous, branched root system. Instead of a straight single root, air pruning creates a resilient root system with lots of root hairs.

## Building Air-Pruned Beds

This can be as complex or as simple as you want to make it. I have made some in 15 minutes that were the size of a large shoe box and others that were 24 feet long by 4 feet wide and took me a few days. Whatever you do, it needs to be strong enough to support the weight of the soil that is suspended above the ground.

You can just build a box, attach hardware cloth to the bottom, and set it up on some rocks or milk crates. The best

This wild carrot pulled from an air-pruned bed dramatically demonstrates the effects of air pruning.

air-pruned bed I made was created by sinking locust posts into the ground, notching the posts, and attaching logs. I laid strong ribs across the frame and screwed hardware cloth down.

## Some Considerations

- Air-pruned beds that are set only a few inches above the ground provide a great hiding place for rodents and chipmunks, who can then steal all your seed nuts. I have found that the higher the bed is, the less cover rodents have.
- Air-pruned beds require more water than trees grown in the ground. They dry out much faster. The soil also freezes faster and warms up faster.
- Trees are very easy to dig up at the end of the growing season compared with those grown in the ground. With the air-pruned beds, I simply pull up trees by their base once they have reached dormancy. Do not try to overwinter trees in the air-pruned beds. The roots will be exposed to huge fluctuations in temperature and significant freezing. I pull up all the trees once they lose their leaves and store them in the ground for the winter if they cannot be planted that fall.
- The long-term effects on trees started in air-pruned beds are largely unknown at this point. What I do know is that the act of making a

Air-pruned beds. The one on the left is constructed on a locust frame with posts sunk into the ground. On the right are 4-by-8-foot frames set on top of milk crates.

taproot is a seedling phenomenon. Once a taprooted species becomes established after a couple of years, it will focus its efforts on lateral roots that are seeking the richest soils found along the topmost layer of earth.
- Air pruning is just one option. You can also raise taprooted species for transplanting by growing them in deep soils that are easy to dig. And of course, you can also direct-seed nuts into their final planting spot and avoid transplanting altogether. This requires a good plan for rodent protection, weed control, and possibly watering.

### Species That Benefit from Air Pruning

This is a short list only based on my experiences. It is by no means comprehensive. So far, I have seen significant improvements in root mass and top growth with all types of walnuts and hickories. Pawpaws and nut pines have also done extremely well in my air-pruned beds. Chestnuts, hazelnuts, apples, and persimmons have been of similar size or less to those grown in the ground. I have really been amazed by the difference between any type of walnut grown in an air-pruning system versus the ground. The trees have been over twice as tall with huge, fibrous root systems.

Air pruning is one method tree growers can use to help our plants through the act of transplanting. It creates trees with fibrous, resilient root systems. This is one method that has worked well for me. Whatever you do with the information here, I hope you find deep satisfaction growing trees from seed. There is almost nothing like it.

## Protecting Seed Nuts from Predators

If you place some nuts or the pits of stone fruits on the ground and walk away, it is very likely that they will get eaten by rodents or birds. Just notice how many acorns an oak will make to reproduce itself. Most of them get eaten. In order to raise nut trees from seed, you will have to offer some protection—sometimes a lot of protection.

Seeds can get eaten, but also as seedlings sprout, critters can pull them up to get the attached nut. This can happen through the entire first growing season and occasionally in the second.

I have a very broad approach at my nursery, which is home to chipmunks, a few squirrels, blue jays, crows, and many mice and shrews. I use many things in conjunction to protect from all these predators.

## Mind-Set

It can be really frustrating to see beds of seedlings destroyed by rodents or birds. Hundreds of trees are destroyed for just a couple of quick meals. It takes a lot of effort, time, and determination to keep seedlings from being ravaged. In order to keep myself on task with this important job, I take on the mind-set that I am the Protector of the Trees here. It may sound corny, but it helps me do this difficult job. I check on my nut seedling beds daily. Once damage starts, you need to take immediate action. If a problem is corrected right away, then trees can be saved. If you don't notice for a week or two, then you may lose everything. To really protect my trees from predators, I am willing to do whatever it takes.

## Bird Repellents

Blue jays are the most aggressive predators of seed nuts in my nursery. To defend against them, I use bird netting over most beds. They will find any small hole in the netting and test it repeatedly once they know what is in there. Netting is a pain to work around and under, but essential if birds know what you have. I have had somewhat decent success using bird scare tape—a shiny metallic ribbon that flutters in the wind and reflects light in the sun. It does keep blue jays out of beds for the most part. It's pretty cheap and easy to install. I have also used a Bird-X noise machine. This runs on a random timer and emits horrible sounds of bird alarm calls. It is pretty annoying to have around, and it is expensive, but it does keep birds from feeling comfortable there. Using a combination of netting, bird scare tape, and the Bird-X mostly keeps the blue jays out. However, once they know what they are after, they will not stop testing the system; adjustments need to be constantly made.

## Traps and Poison

If you put out poison, you are starting a chain. The poison that is ingested by an animal or bird who dies does not go away. Scavengers who eat the poisoned critter can also become poisoned—and still the poison will be in them. I personally find this unacceptable. I do use traps, though. The squirrel population is low enough in my nursery that I can manage them by trapping with Havahart traps. I bait it with chestnuts, walnuts, or whatever they are after.

For mice and chipmunks, Havahart traps are not realistic because the critters are too numerous. I use snap traps for them. It can be disheartening to kill these creatures if you are not eating them. I think about how many trees I am protecting. Each nut tree I am able to establish in the world will feed many generations of mice and chipmunks. I know that doesn't help the individual who gets killed, but it helps me wrap my head around the practice. Generally, I set traps out in the beds and will kill several mice and chipmunks for a week or two until they stop coming. They will come back at some point, but they can really be kept at bay with snap traps.

## Breaking Off the Nut

If you are not raising a lot of trees, you can start them safely indoors in old milk cartons or something similar. Once the trees are 6 to 12 inches tall, reach down and break off the nut from the seedling. You can then plant this tree outside.

You can also break the nut off trees that you start outside at any point. Just be careful not to pull up the whole tree. Unfortunately, once critters are used to digging up seedlings to eat nuts, they often continue to pull up seedlings even if no nut is attached.

## Using Air-Pruned Beds

I think that this may be the most effective way to protect seedlings. The beds need to have hardware cloth well secured on all sides and bottom. Building two beds side by side allows you to make a hoophouse around them, using their outside walls as the sides of the house. The hoops are covered in chicken wire or hardware cloth. This is a lot of infrastructure, but it will last year after year. Mice and chipmunks cannot get in from below, and squirrels and birds cannot access the beds from above. In the near future I will probably be starting all of my chestnuts and hazelnuts this way.

CHAPTER FIVE

# Propagation by Grafting

People have been grafting trees for thousands of years. Grafting is a valuable propagation skill to have. You can clone trees that are difficult to root, and you can change mature trees over to a better variety. There is a lot of mystique around grafting, and many folks are intimidated to start. The truth is that grafting takes practice, but it's not that hard.

Grafting means a stem or bud is joined to another plant with roots. The two fuse together. The way it works is through callusing. When we make a cut on a tree, it will form callous tissue if it's healing properly. Callous tissue is very sensitive. When you have two branches pressed against each other forming callous tissue at the place where they meet, they will form a graft union. You can increase the surface area where the callous tissues will meet up by making specific cuts.

There are a huge number of grafting cuts and techniques. They all follow the same principle: *Line up the cambium layer so that callous tissues line up.* The cambium is a thin layer of cells between the inner bark and the wood. Use very sharp tools to create the cleanest possible cuts. This will not only allow for better healing of the stem, but also increase the contact between the two cuts.

A grafting knife is exactly what it sounds like, and if you're serious about grafting more than a handful of trees you should get one. Keep it extremely sharp. I touch up the blade on mine every time I sit down to graft. Some people use box cutters. And some growers use grafting tools; these tools make an omega cut that eliminates the skill required of a knife wielder. I've never used a grafting tool or a box cutter, but I have used grafting knives a lot. They are small, simple tools that work

## Propagation by Grafting

very well and can last a lifetime. A normal grafting knife is beveled on one side. Unfortunately for me and all other left-handed people, they are beveled for right-handed people to use. If you're a lefty, then you can get a budding knife, which is beveled on both sides, or you can grind a grafting knife to be beveled on the opposite side.

Grafting knives work very well, but take practice. There are endless YouTube videos of people grafting, including my own. Many garden clubs and cooperative extensions offer grafting classes. It's a good idea to find someone who knows how to graft and watch them. Before you slice open any trees, practice on green stems of the same species. I sliced through hundreds of practice sticks until I felt comfortable cutting any of my own trees.

Be careful. There are few grafters out there who have not cut themselves at some point. The last time I cut myself I had to go to the hospital for stitches. Even though it's not cool, I wear a safety glove now. You can get knifeproof gloves that are used by chefs for a few dollars. It is worth not missing a few weeks of work or damaging your hands. I always wear a glove on my non-knife hand. I wrap my knife hand thumb with thick tape. All of the other grafters I know do not wear a glove. I'm a dork, but it's worth it. Some people do wrap their fingers with tape, which looks cool and will stop minor cuts. I had tape on my finger when I sliced it open. The only cut that is really dangerous is the tongue cut on a whip-and-tongue graft. You can sometimes cut all the way through the scion unexpectedly. Grafting knives are wicked sharp.

### Scionwood

The scion is the top of the grafted tree. It is what you are splicing onto the rootstock. The best scions are taken from vigorous, straight shoots. You will find the best scionwood near the tops of trees, from water sprouts growing off the trunk,

Scionwood bundled up and ready to be stored for winter bench grafting.

and from trees that are heavily pruned. Most commercial nurseries maintain scion trees that have the entire top cut off every year. The resulting growth makes perfect straight shoots.

Scions are collected while they are dormant. I collect mine in mid- to late winter on mild days when the sap is rising. I store them the same way that I store cuttings: either in bags in the fridge with moist paper towels around their stems, or in the basement packed in sand.

## Rootstocks

The rootstock will influence the growth habits of the grafted tree. There are full-sized (standard) rootstocks and dwarfing rootstocks. They are either clonally propagated or grown from seed, or they are established trees. Rootstocks have to be compatible with the scion. Most species match up with themselves only. There are a few exceptions, but not many. You can graft pears onto hawthorn, or quince onto pear, but for the most part you've got to keep it within the same genus. Apples go on apples. You can't just graft a branch of apple onto an oak tree and expect it to survive.

It's nice if the rootstocks are the same diameter as the scionwood, but this is difficult to achieve and is not essential. What you don't want is scionwood that's thicker than the rootstock. This will form an ugly graft union that is not very strong.

## Bench Grafting Versus Field Grafting

Bench grafting is done indoors or at a table outside in the shade, and involves grafting small trees that are bareroot. The trees are picked up, sliced, spliced, and placed in a bucket of water until they are heeled in or planted. Field-grafted trees are grafted to rootstocks that are in the ground outside.

Bench grafting allows the grower to work inside during cold weather. Field grafting has to be timed for specific weather depending on species and technique.

## Lining Things Up: Grafting Techniques

In all the following techniques, it's important that your cuts be decisive. One slice makes one cut. You are not whittling. One slice pressed against one slice allows for the cambium layers to line up without any bumps or dips disrupting the connection. If your scion is smaller in diameter than the rootstocks, then just line them up on one side.

## Whip and Tongue

To make a whip-and-tongue graft, start by making a slanted cut with a steep angle. About a third of the way down from the tip of this cut, slice down until you cut about halfway down the slanted cut. Repeat this on the scion. Slide them into each other. You may need to push somewhat firmly.

Whip-and-tongue graft. Notice how the two stems lock together.

Whip and tongue is the most difficult cut to make, but it's the best and strongest in my opinion. The stock and scion are joined with maximized contact. The unions formed are very sturdy right away; some other unions will take years to strengthen.

Whip and tongue makes sense for bench grafting. It's not easy to make the cut when rootstocks are in the ground, though it is possible. With practice whip and tongue is easy. After a while, you won't even think about it and will be able to graft hundreds of trees per day with this technique.

## Cleft Grafting

This is pretty easy to learn and does not take nearly as much practice as whip and tongue. The union is pretty ugly for at least a few years and is not nearly as strong initially. Over time, though, it will grow stronger and stronger.

In cleft grafting, split the rootstock down the center with a sharp knife. Don't push on the end of a knife with a bare hand; use a wooden mallet to tap the knife down if you need to. Cut your scion with two steeply slanted cuts, forming an even bevel. The end of the scion will be the same shape as a flathead screwdriver. Stick it into the split rootstock, pushing it down as far as you can. Be sure to line it up on

Bark graft healed over.

### Bark Grafting

one side, so there is contact between the cambiums.

I know I'll annoy someone here, but I think bark grafting sucks and is a waste of time. I only mention it in case you are reading about it somewhere else and want to use it. In bark grafting, beveled scions are inserted into a cut branch or trunk between the bark and wood. These grafts are really weak and break off easily if a bird lands on them or the wind is too strong. They also are made on such a big wound on the rootstock that they usually don't heal well. That's my two cents. I'm sure others have had better experiences.

# Budding

Budding is the insertion of a single bud into a rootstock. Generally trees are budded in late summer. The bud is left in place and remains dormant until the following spring. The top of the rootstock is cut off just above the bud in the spring, and all the new growth comes out of that single bud. Budding works nicely when you want to graft trees actively growing in a nursery row. It is usually not used for dormant, bareroot stock.

### Chip Budding

Chip budding is pretty easy to learn. Cut at an angle on the rootstock about a third or a quarter of the way into the trunk. Make a same-angled cut about half an inch above. The second cut will meet the first one and a chip of wood will fall out. Do the same thing to the scion, except do it around a bud and keep the chip of wood that falls out. Insert the chip with the bud into the rootstock. Make sure at least one side is well lined up. You can tape right over the bud with grafting tape.

## Propagation by Grafting

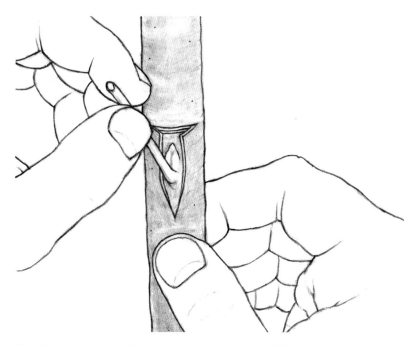

T-bud being inserted into rootstock. Illustration by John Walsh.

### T-Budding

This technique is easy enough, but must be done when the sap is running in the trees. Usually July and August are best. It helps to heavily water trees during the week prior to T-budding; the bark will peel more easily. Slice a T-shape into the rootstock. You can open up the T-cut with the tip of the knife. Slice a bud off a scion. Cut the leaf off, but leave the leaf stem. It makes a nice handle. Try to only cut the bud off without going deeper than the inner bark. It will naturally want to separate from the wood if you pay close attention.

Peel back the bark on the rootstock with the tip of a knife, insert the bud into the T-cut, and wrap it up. It will insert more easily if the bottom of the bud piece is slightly pointed.

## Taping and Wax

Taping up grafting cuts is necessary. Left on their own, the unions will be prone to drying out. The tape will also help hold things together. To seal up grafting unions in the old days, growers used string and compounds made with pitch, wax, and manure. Grafting tape is a lot easier. It is the same thing as surgical tape. It stretches and sticks to itself.

Wrap unions tightly, stretching the tape as you go. Be careful not to knock your scion out of alignment when you wrap it. After the union is wrapped, cut the scionwood so only one or two buds are left. If you leave the scion too long, it will be top-heavy and not as strong. It will also draw a lot of water when it leafs out. One nice bud is really enough. It's okay if you tape over a bud. They can grow right through tape or wax.

Cover the end of the scion with beeswax or grafting tar. Wax needs to be melted, which is cumbersome in the field but no bother during bench grafting inside on a winter day. The tar can be carried into the field easily, but it's pretty gross stuff.

## Storing

After being wrapped and waxed, the bench-grafted tree is ready for storage. Most of my grafting is done in February and March when the ground is covered in snow. I store my bench grafts by packing them into either sand or very old sawdust. They can be packed in bins or pots. I store mine in the unheated basement.

## Weather

Different trees callus at different temperatures. Very few are able to callus at subfreezing temps. Apples and pears can handle being stored in the 30 to 40°F (0–5°C) range and callus well. Most other trees need warmer temperatures. Walnuts will have trouble callusing below 70°F (21°C). This will influence timing and grafting methods. For example, peaches are budded in summertime. I bench-graft mulberries in May and keep them safe from frost.

I plant out all of my bench grafts after danger of hard frost. I've had the temperature plunge into the low 20s (−7 to −2°C) after planting out apple grafts. They showed no sign of damage, but I might have just been lucky. It might be worth experimenting to see what you can get away with. In general, though, keep fresh grafts safe from cold weather.

## Propagation by Grafting

# Aftercare of Grafts

After the grafted tree is planted and growing, it needs some attention. Every few weeks during the growing season, check for sprouts coming off the rootstock. Wipe these off with your fingers while they are young and tender. This will force the tree's energy into the scion. In most cases the scion will take over after a while, and the tree will fully accept it as the top. You still want to remove any root suckers over the years.

# Top-Working Older Trees

It's pretty amazing to look at an old apple tree with terrible fruit and convert it into a tree with 30 varieties of your choice. Each branch can be a separate variety, or you can have them all be the same. It's tough to keep track of multiple grafts in the same tree over time. It takes quite a bit of pruning and attention, but the reward can also be high.

An older apple tree top-worked by grafting onto its younger branches. The union is quite strong. These stems were cleft-grafted two years prior to this photo. Notice the healed scar on the right branch.

A graft on a mature tree usually starts bearing about three years after it was made (at least in the case of apples).

Generally, cleft grafting is the preferred method for splicing into branches thicker than a pencil and smaller than 6 inches in diameter. It's very difficult to split and hold open larger branches. People use bark grafting on larger limbs. (Personally, though, I think bark grafts on big limbs make poor unions.)

In my experience, the best way to top-work bigger trees is to pick some sucker shoots that are one to three years old and either cleft-graft them or use the whip-and-tongue technique. You can then cut back the rest of the trunk. Some people leave a nurse limb or part of the original top in place for a year or two until the graft is strong.

The growth powered by an old root system concentrated on a small branch can sometimes be phenomenal.

Some trees are too old and worn out to rejuvenate and top-work. These ancient ones often will not heal well from big wounds. You'll know a tree is past time for pruning and grafting when it has big holes or large dead sections in the trunk. I have learned to leave these ones alone and just enjoy their character.

CHAPTER SIX

# Propagation by Layering

*L*ayering is a passive form of propagation. With certain varieties it happens all on its own. There are many different layering techniques, but they all follow the same fundamental concept: A stem is encouraged to form its own roots while it is still attached to the plant.

A branch is bent to the ground and weighed down with soil and stone until it roots. The stem is propped back upright with another stone. Rooting times vary depending on species and conditions, from months to years.

## Simple Layering

Here a low branch is bent down to the ground. You can place a stone on top to hold it in place. For many species that are easy to root, this is enough. For more difficult species, you can tie a twist-tie wire around the stem. The wire will restrict growth as the stem expands through the growing season. The branch after the wire will feel the stress and be encouraged to throw down its own roots. You can also wound the stem by scraping it where it will be in contact with soil. This wounding will encourage callous tissue to form. Some growers dust the wound with rooting hormone.

## Mound Layering

Instead of bending down a single branch, in mound layering you pile dirt or old sawdust over the whole bush. By the way, you can make a tree into a bush or encourage a bush to make more stems by cutting the whole thing to the ground in the winter.

Create a mound layer by simply adding a growing medium all around the stems. After rooting, pull the medium back and clip out the rooted stems for replanting.

In mound layering you should replenish the growing medium around the stems at least a couple of times throughout the season, as it will settle quite a bit.

## Tip Layering

This is where just the tip of a stem is buried. Certain varieties naturally tip-root, and these are the ones this technique is best for. You can encourage more tip rooting than the plant would naturally make by weighing the tips down with stones. You can also allow grass to grow fairly thick around tip-layering plants. The grass will help the stems stay in place on windy days or when brushed by.

You can encourage a plant to make more stems for tip layering by pinching back the tips of stems in early summer. This will cause the stem to branch out, and each branch can create another rooted plant.

Some species that naturally tip-root are black raspberries, blackberries, gooseberries, and forsythia.

## Serpentine Layering

This works best on vines and some very fast-growing shrubs. As a layered stem is rooted and continues to grow, the next section of growth is buried, and so on. I have had individual goji berry stems layered six to eight times in a season.

## Air Layering

This is a very time-consuming technique, but is especially useful if you really want to have a certain tree on its own roots. It makes sense to have certain trees on their own roots and not grafted for lots of reasons, but especially if you want to build stool beds with them.

To air-layer, bring the soil up to a branch that is too high to bend to the ground. Wound the stem heavily; many growers peel off an entire section of bark. Take a plastic bread bag and cut the bottom out so that you have a sleeve of plastic. Slide it over the wounded stem. Tie the bottom tightly with a twist-tie. Fill the bag with moist growing medium. Many growers use peat moss, coconut fiber, or old sawdust. Fill the bag completely. Close the top with another twist-tie. Shade the bag by wrapping it in either aluminum foil or burlap. Check back every few weeks to see if roots are pushing against the sides of the bag. Once they are, clip out the stem and you will have a very large rooted plant.

# Stool Layering

Stool beds are where plants are grown just to be layered. They are cut down to the ground every year as layers are harvested. It makes for a simple, effective propagation technique.

There are two methods I use for starting a stool bed. One is to plant trees horizontally. The second is to plant a tree and cut it down in the dormant season to turn it into a bush (this is the same as mound layering, but in this case you are building a whole bed). With either method, your goal is to have an entire bed sending up a dense stand of shoots.

When trees are planted horizontally, the trunk and branches are laid along the ground, just at or below the soil surface. All of their buds will grow skyward. As the stems grow over the summer, pile old sawdust or soil on them until just the tips are showing. I do this three or four times over the summer. The stems will root into the medium you pile on. You can think of this as similar to hilling potatoes. Just keep piling medium onto the stems, leaving the tips exposed. In the fall or the next spring, simply cut out the stem with some roots attached.

After you harvest the rooted layers off a stool bed, feed the soil with compost or some nutrient-rich mulch. As the years go on, stool beds

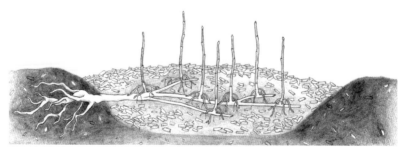

Planting a young tree horizontally causes many leaf buds to turn into vertical shoots. Each of these vertical shoots is mounded with sawdust. This is a fast way to start a stool bed. Illustration by John Walsh.

## Propagation by Layering

develop amazing soils and excellent growth. The roots of the mother trees just get bigger and more mature, while the soil is constantly fed by incessant mulching.

Most commercial apple rootstocks are grown in stool-layered beds. Millions of trees are produced this way at a rate of around 60,000 rooted stems per acre.

## Harvesting Layers

You can harvest rooted stems anytime. I like to let them get quite strong before I cut and dig them out. Sometimes I leave stems to root for two or three years before separating. The resulting plants can be huge.

Some plants take only a month or so to form roots, but others require two years, and still others may never do it. It's fine if it takes a while. Layering requires almost no energy on your part once you've started it.

It is preferable to harvest layers while they are dormant plants. A pair of pruners or loppers and sometimes a shovel are all that's needed.

All my stock plants for layering grow stronger every year and send up bigger and more numerous stems each season. They are fed generous helpings of compost and mulch. Layering is slower than using cuttings, but also much easier. Watering needs are near zero with all the mulch. I think layering is one of the most overlooked, easiest ways to propagate lots of plants. If you do it intensively, it can be quite productive and profitable.

# PART TWO

# The Allies

I am humbled to be writing about the following trees. They are incredible beings, and I hope that I may represent them with the respect and reverence they deserve. Speaking about them is like talking about old friends. It's not that easy to tell a stranger about someone you have known for years. I will do my best, but my real goal is for you to meet and know these trees yourself.

These trees are our allies. They do work that we need done—making food or medicine, carbon harvesting, feeding wildlife—all the while living off sunlight and nourishing our souls with their awe-inspiring presence. For the sake of book length, I have selected only 10 genera here. I could have easily chosen dozens of others. All trees have a unique story to tell and have their own specific contributions to people and wildlife. These trees are some of my favorites, but there are many more to learn about.

I definitely don't know everything about these trees. Everyone has their own perspectives and knowledge. I can only hope to share mine here. Please understand that there is much more to learn. I find that I learn new things about these trees every year that I work with them. I imagine this will continue as long as my partnership with them does.

CHAPTER SEVEN

# Chestnut
## The Bread Tree

There are legends of a time when manna fell from the sky, and people could just pick up all the food they needed. Mature chestnut trees are pretty close to making that story a modern-day reality. Chestnuts can live for thousands of years and rain down dependable annual crops. They are magnets for wildlife, staples in several cultures, as magnificent as an oak, provide quality, durable timber, and are real epicenters of life.

Chestnuts are in the Fagaceae family, the same as oak and beech. These are the sturdy old giants of ancient forests from Maine to Turkey to Japan.

## The *Castanea* Genus

There are several species of chestnut stretching around the temperate world with relatives in the tropics. These trees are not that well known in the United States, but they are beloved, widely used, and very common trees in many other countries. Each of these species has its own unique traits, strengths, and challenges. The members of the *Castanea* genus are some of the most generous and productive trees on Earth.

### American Chestnut (*Castanea dentata*)

This tree dominated the forests of the eastern United States until it was devastated by an introduced fungus from Asia. American chestnut is a fast-growing tree with bar-none timber form. It bears small, very sweet nuts that peel easily. It is the most cold-hardy of the chestnuts. More on the American chestnut in a few pages.

Allegheny chinquapin loaded with small, sweet nuts.

## Allegheny Chinquapin (*Castanea pumila*)

This is a shrub occasionally reaching up to 25 feet tall. It is native from Pennsylvania south through Appalachia. Chinquapins bear one nut to a burr, as opposed to the usual three found in most chestnuts. Chinquapin nuts are very small, about the size of a blueberry. They are by far the sweetest of any chestnut. Allegheny chinquapin is susceptible to chestnut blight, but it is able to send up new shoots when old ones die of blight. It can live a normal, healthy life even with blight present. Allegheny chinquapins are widely planted for turkeys, grouse, and other wildlife. They have been used in various breeding programs to add their highly sweet flavor to trees that bear bigger nuts.

## Ozark Chinquapin (*Castanea ozarkensis*)

Botanists used to recognize this as an individual species, but they now say that it is the same as *C. pumila*. I'm not a botanist, but it seems obvious to me that the Ozark chinquapin is a distinct species. The trees do have similar nuts: small, sweet, and one to a burr. However, Ozark chinquapin is a forest tree growing 80 feet tall on the limestone soils of Arkansas, while Allegheny chinquapin is a medium-sized shrub of Appalachia. It seems fitting that the Allegheny chinquapin would be named *pumila*, meaning "dwarf." But that name shouldn't apply to this unique 80-foot-tall tree of the Ozark Mountains.

Ozark chinquapin is very blight-susceptible, and few have survived to this day that are not old stump sprouts dying back to the roots continuously. It has not received the attention that the American chestnut has, but there are a few people growing and breeding them. Sandra Anagnostakis of the Connecticut Agricultural Experiment Station grows many *ozarkensis* hybrids.

Ozark chinquapin nuts are 15 percent protein—a much higher ratio than found in any other species of *Castanea*.

Ozark chinquapin is the only species resistant to Asian chestnut gall wasp,[1] which is a significant pest. This species offers another block in the wall of *Castanea*'s genetic resiliency.

### European Chestnut (*Castanea sativa*)

Cultivated throughout Europe for thousands of years. Originating from the Mediterranean east to Iran, *C. sativa* was spread throughout Europe centuries ago by the Romans and other travelers who carried seeds with them. Also known as the sweet chestnut, *C. sativa* reigns over the continent from England to Asia Minor. It is a massive forest tree found in large, pure stands, as well as in orchards, in parks, and along the streets of Paris. There are productive orchards of European chestnut that are over 1,000 years old in Italy, Corsica, Turkey, Azerbaijan, and several other places where mountains are covered in ancient forests of chestnut. The nuts are gathered by hand in most regions, as people have done for centuries. There is one tree in Sicily that is known as the hundred-horse chestnut. Sometime in the 1500s a princess was traveling with her entourage when they were forced to take cover from a thunderstorm under this tree. Apparently she and a hundred horsemen could all fit under the massive canopy. This tree, growing at the base of Mount Etna, is estimated to be 4,000 years old. It has a trunk that measures 190 feet in circumference.

European chestnuts can produce some of the largest nuts. Italian *marrones* are famous for their size and flavor. European chestnuts have an enormous cultural heritage. They are found in many recipes, from candies to breads to stuffings. However, European chestnuts are very hard to peel, as the pellicle (skin under the shell) clings to the meat tenaciously. European chestnuts are also highly blight-susceptible.

Chestnut blight was accidentally introduced into Europe in the 1940s; it was believed that the trees would be wiped out just as American chestnuts were. Strangely, the disease was halted in its tracks by a wild virus. The virus has kept the blight from decimating ancient European groves, but the blight has still caused significant damage. This virus has not been as successful as in the US because we have a different strain of blight here. There are actually hundreds of strains of blight and many different viruses. Today orchards in Europe are being planted with Japanese × European hybrids for disease resistance.

Asia also is home to several unique species. There is a long tradition of using chestnuts in Korea, Japan, and China. People eat more chestnuts in Asia than anywhere else in the world. China grows more than 10 times as many chestnuts as the next country in the world, which is Korea.[2]

### Japanese Chestnut (*Castanea crenata*)
Often cultivated as a spreading tree with very large nuts and good adaptability to adverse soil conditions. Japanese chestnut has the highest level of resistance to chestnut blight and phytophthora root rot (ink disease). Reportedly the nuts lack flavor, but the ones I have tasted were good. This tree is currently being used to breed super-producing trees with enormous nuts.

### Chinese Chestnut (*Castanea mollissima*)
A tree of high variability that makes up forests and orchards, the Chinese chestnut has been the most widely planted exotic species of chestnut in

# Chestnut

Japanese chestnut in Caldwell County, North Carolina. Photo courtesy of Paul Sisco

the United States. The trees are highly blight-resistant. Hardiness can range quite a bit within the species. Most Chinese chestnuts in the US were brought here from sources in southern China, with a climate similar to zone 8 or 9. There are Chinese chestnuts that live at more northern latitudes all the way up into Manchuria, where it gets very cold. The Chinese chestnuts we find in the US generally lack in timber form. However, there are forests of Chinese chestnuts in China with very large trees.

## Seguin Chestnut (*Castanea seguinii*)

This is a medium-to-small forest tree from central and southern China. It bears small nuts, three to a burr. Seguin chestnuts often have the odd habit of continuously flowering all summer until fall. They are extremely precocious, sometimes bearing nuts by age two. They have been used primarily in wildlife plantings. I believe they hold great potential for breeding extremely high-yielding trees. They are hardy to zone 5 in some populations.

## Henry Chestnut (*Castanea henryi*)

This tree is native to southwestern China, where it grows to heights of around 80 feet. It bears a single small nut to a burr. It has not been used in the US, because it's not that cold-hardy. An interesting species, nonetheless. I would love to walk in a forest of Henry chestnuts. They often grow in mixed forests of Seguin and Chinese chestnuts.

### *Castanopsis* Genus

This genus is very closely related to chestnuts and includes 120 species. Almost all members of the genus are native to tropical and subtropical regions of Southeast Asia from Borneo to Japan. There is one species native to California, *C. chrysophylla*, known as the golden chinkapin, that grows among the redwoods. Castanopsis trees are large timber trees growing in a wide range of soils, from bog to ridgeline. They make edible nuts very similar to small chestnuts.

Around the world chestnuts grow in many different conditions and in several forms. They can be found in both acid and alkaline soils, cold and hot regions, and humid and dry areas. Chestnuts can be massive forest trees, small spreading orchard trees, or tough thicket-forming shrubs. There is a chestnut tree well adapted to almost every bioregion in the world.

## The Epic Saga of the American Chestnut

The American chestnut may well be the greatest and most useful forest tree to ever grow on this Earth. Its decline is considered by many ecologists to be one of the greatest ecological disasters to strike the US since European contact.

It is hard for us today to understand what was lost because we did not witness it. Imagine working in your yard and noticing an apple tree with wilted leaves. A few weeks later, the tree dies. You're sad about this and tell a friend, who tells you that they had the same thing happen. Then you hear it from lots of people. It's on the news. Apple trees are dying, orchards are wiped out, wild trees disappear. No one knows what to do. Before you know it there are no cider barns, no crisp fruits to sink your teeth into, no apple blossoms in the spring, no fruit in the supermarket. How would you feel? As the years go on, you might try to explain to young people what an apple tasted like, what it felt like to bite into apples; you might describe the trees' gnarly growth habits or the smell of cider in the barn or the taste of applesauce. They would never understand. The apple tree would be gone and life would go on. Other trees would be there, but none would be the apple. This is basically what happened to the American chestnut. The chestnut was no less loved or used than apples are today. It was a tree with full cultural, economic, and wild significance. We are the people who were born after its loss. All we have are the stories and a handful of pictures to go by.

American chestnuts growing in the Appalachian Mountains. Photo courtesy of Forest History Society Durham, North Carolina.

*Castanea dentata* dominated the eastern US, making up roughly one-fourth of the trees in its range. This is a huge percentage, considering the diversity found in the eastern deciduous forests. Even maples, oaks, and ash are not that common.

American chestnut trunks were massive, often 10 feet or more in diameter, with canopies reaching 130 feet in the air. These arrow-straight, towering trunks were made of high-quality, rot-resistant timber. The wood was used for barn beams, house framing, furniture, telegraph poles, fence posts, paper pulp, caskets, and cradles. There is no wood so versatile as American chestnut. It has the durability of black locust, the straight grain and splittability of ash; it's as stable and easy to work as pine, and very fast growing.

The wood value alone would have made the American chestnut a highly valuable tree. Adding the dependable crops of nuts makes this tree stand alone in its excellence. The wildlife value of American chestnut was unparalleled, as nuts could fill the forest floor more than a foot deep in some years. Along with wildlife, people also ate wild American chestnuts. They grazed their animals under these magnanimous giants during the fall and gathered nuts by hand. Chestnuts were collected in great quantities throughout the Appalachian Mountains, and roasted and sold on the streets of towns like Boston, Philadelphia, and New York. Train-car loads were filled with this wild crop. Today many families find financial relief with their end-of-year tax credit. Back then people found their Christmas bonus in the form of selling what chestnuts they could gather in the mountains.

The American chestnut was a keystone species in the ecology of the Appalachians. It was culturally fixed in the minds of Americans, and used widely. Tanneries cranked out leather that was processed with the tannins of chestnut bark, paper mills pulped the wood, railroad companies laid track with timbers, people built barns and houses, fences, and chairs. They ate the nuts raw and roasted every fall. And then it all crashed.

In 1904 chestnut blight, *Cryphonectria parasitica*, was discovered in the Brooklyn Botanic Gardens. From there it spread like a wildfire, consuming trees and turning forests of green into silvery gray ghost woods. Within just 25 years an estimated four billion trees died. An entire ecology, an entire culture, was wiped out. While the trees were dying, the US Forest Service advised people to have all their chestnuts logged. Believing there was no hope, they told folks to get some lumber out of it while they still could. We will never know how many resistant trees were killed in this shortsighted practice. Sadly, this mind-set persists today, as foresters commonly advise landowners to log all their ash and hemlock trees before the coming crash.

The Shelton family in Tennessee pose in front of a blight-killed American chestnut (1920). Photo courtesy of the Great Smoky Mountains National Park Library.

*Cryphonectria parasitica* is a fungus whose spores spread by wind. Its origins lie in Asia, where trees there have co-evolved with the fungus. When Japanese chestnut seedlings were brought over to the US for people's gardens, no one noticed that these seemingly healthy trees carried the blight with them. The American chestnut had never encountered this fungus and so had almost zero resistance. People scrambled to save the chestnut tree in vain, employing all sorts of strange strategies over the next several decades before giving up for the most part.

There were some really wonderful early attempts at saving the American chestnut, notably the work of Arthur Graves. He planted several thousand seeds of anything he could get his hands on, including every species of chestnut from around the world. He crossed every species and then crossed the resulting hybrids. Many of his trees are still alive and maintained to this day at the Connecticut Agricultural Experiment Station by the committed and innovative work of Dr. Sandra Anagnostakis. Graves was never able to find the winning combination of a true timber-type tree and full blight resistance—though he found many trees that came close. The work of Graves and Anagnostakis continues today. Dr. Anagnostakis continues to plant trees, make controlled crosses, and spread hypovirulence. Today the Connecticut Agricultural Station and Sleeping Giant State Park are home to the largest repository of chestnut genetics in the world.

The forest left behind by Arthur Graves at Sleeping Giant State Park, in Connecticut. All the trees in this photo are mixed species of chestnut.

Other attempts at saving the American chestnut included cutting out all the trees in a large swath of land across Pennsylvania to act as a kind of firebreak. The budget for this project was enormous but could not keep up with the winds that carried blight spores. In the 1950s, when nuclear power was at the height of popularity, people irradiated nuts to hopefully invoke a mutation. Of course, this failed miserably.

The breeding program at the Lockwood, Connecticut, Agricultural Experiment Station run by Dr. Sandra Anagnostakis. This program includes species from all over the world and extends through many different plantings. This particular planting is a mix of American chestnut and Ozark chinquapin and also includes genetics of Japanese and Henry chestnut.

As the years wore on, and generations passed, interest in the chestnut grew less and less. The American chestnut became a legend, with little practical hope of recovery. That is, until two visionary men in the 1970s, Dr. Charles Burnham and Phil Rutter, came up with a plan. It was simple enough: Cross American trees with resistant Chinese trees. The resultant seedlings would then be backcrossed with American trees again and again until they had a tree that would behave like the American chestnut of ancient forests. The tree they worked to create would have fifteen-sixteenths American genetics. This would take several generations of crosses. Dr. Burnham knew he would never live to see the full breeding program to its completion, but he started it nevertheless as a selfless act. This was the beginning of The American Chestnut Foundation (TACF).

TACF would go on to plant thousands and thousands of seeds from multiple lines of genetics. For every tree they leave to grow,

# Chestnut

Chestnuts at Hemlock Grove Farm, West Danby, New York. These trees were planted by Brian Caldwell, who has been planting and selecting productive, blight-resistant trees for the past 40-plus years. Several of Brian's trees originate from the plantings at Sleeping Giant State Park.

hundreds are cut down in this rigorous and highly organized breeding program. Today TACF has trees that are fifteen-sixteenths American and display both timber form and blight resistance. These trees are being planted experimentally in parks, at private homes, at institutions, and in reforestation efforts.

Many folks, including myself, continue to plant and grow 100 percent pure American trees as well as hybrids. American chestnut seed is available through TACF state chapters. The trees grow quickly. They typically live for 15 years before succumbing to blight. In this time they produce small crops of nuts and excellent pole wood. Since the blight cannot kill the root system, the trees sprout back after the blight knocks them down. They can be kept going indefinitely in a coppice system. Growing American chestnuts from seed also expands the genetics of this magnificent species.

There are other programs in action today. The American Chestnut Cooperators' Foundation grows only 100 percent pure American chestnuts. They maintain trees that exhibit resistance through grafting and seed collection.

SUNY's College of Environmental Science and Forestry, along with the New York chapter of TACF, is genetically engineering blight-resistant chestnuts. They have introduced a wheat gene to resist the fungus. The trees are currently not released for public planting pending government approval. This program is considered by some native-plant enthusiasts to be superior to the traditional breeding program of TACF because the trees are less "contaminated" with Chinese genetics. However, it's interesting to note that Chinese chestnut and American chestnut share most of each other's genes; the hybrids that TACF grows are actually over 99 percent genetically American chestnut.

The story of the American chestnut is far from over. Today we can grow resistant hybrids or pure American trees. We can bring this species back into our parks, homes, and wildlands. There really is no reason not to. Millions and millions of people live in the range of the American chestnut. If just 1 percent of them chose to plant a few trees, we'd have a lot of chestnut trees around.[3]

## Wildlife Value

What chestnut trees provide to wild animals is truly unparalleled. The only tree that comes even close is the oak, but oaks are very different in their fruiting habits, with a profound impact. There are many species of oak found around the world, very few of which bear regular annual crops. Here in the northeastern United States, oaks flower in early May. At this time we have highly fluctuating temperatures, and it is not uncommon for a freeze to wipe out nearly all the oak flowers. If no frost destroys the acorn crop and the oaks fruit heavily, they will often take the next year off and produce only a very light crop, even in the absence of late frosts. Generally, oaks produce a heavy crop every two or three years. In fact, almost all trees produce heavy seed crops on an irregular basis. The chestnut stands alone in reliability. It is the manna tree.

Chestnuts flower very late here, around the end of June or early July. Their blossoms almost never encounter frost. They bear fairly consistent crops every single year. This is a highly valuable and unique trait among fruit and nut trees.

# Chestnut

Chestnuts and acorns are heavy in starch. They are staples not just in the diets of humans, but also in the diets of many species of wild animals. They are the equivalent of rice or potatoes falling out of the sky. If you are a chipmunk or a squirrel who depends on this, then missing a crop every other year is totally devastating. This is why many rodent species (and their predators) in forests of oak, hickory, and beech will have boom and bust cycles in their populations. The addition of chestnut trees to a forest can significantly temper fluctuations in populations.

The list of wild critters that eat chestnuts is indeed a long one. It includes bears, wild turkeys, deer, raccoons, jays, opossums, skunks, foxes, squirrels, chipmunks, and several other rodent species. While it is easy to look unfavorably upon rodents, they are the backbone of the diet of several predator species that we revere, such as owls, hawks, and bobcats.

Chestnut leaves are eaten by over 100 species of Lepidoptera (butterflies and moths). Their flowers are a significant source of nectar for pollinators. Bark, buds, leaves, and twigs are fed on by a plethora of mammals.

It is hard to overestimate just how important chestnut trees can be for wildlife. They create a reliable staple food source that is nutritious, thin-shelled, and easy to consume.

Chestnut trees have a truly magnetic effect upon wild animals. They are centers of activity. Animals, birds, and people will travel from all directions to a stand of chestnuts. I can think of few greater legacies to leave behind than a stand of chestnut trees.

## Wood

The timber value alone makes chestnut trees worth growing. Plantations of pine, spruce, and poplar offer people fast-growing timber resources. Chestnuts are another viable option for this, with many added benefits. It is a beautiful straight-grained hardwood that has found its way into high-value items like furniture, trim, and timbers.

Chestnut trees can grow very fast. They are easily coppiced. This means that they can be cut again and again because they will sprout vigorously from the stump. Coppicing chestnut is an old tradition in the U.K., where trees are put on rotation cycles ranging from 1 to 50 years depending on the desired wood products.

Chestnut wood is highly valued for many reasons. Not only is it a beautiful wood, but it is also light and strong. Chestnut is very stable and resists warping and cracking. It is easy to work and highly rot-resistant (rot resistance varies among species, with *C. dentata* and *C. ozarkensis* having the highest levels). Chestnut wood is a viable alternative to pressure-treated lumber; it is certainly more durable in contact with the soil. The rot resistance of American chestnut is on par with black locust. The main difference between black locust and chestnut is the weight and ease of use. Black locust is the gold standard in durability, but it is really hard to work with. Making things from locust often means broken drill bits, bent screws, and nails that will never be extracted by anything but fire. Chestnut is an easy-to-use, fine-grained wood that also makes durable fence posts. What other wood is this versatile? It splits easily, carves wonderfully, finishes smooth as glass, and lasts a lifetime if left out in the mud.

My friend Steve Serik of Hawk Meadow Farm inoculates shiitake logs for commercial production. He says chestnut logs grow the most and best shiitakes.

The wood also has such a mystique that I believe a person could make a living selling chestnut walking sticks or key chains.

Hybrid chestnuts gathered at Hemlock Grove Farm, West Danby, New York.

## The Nuts

Chestnuts can be eaten so many different ways. The most famous way is roasted, but they can also be steamed or boiled, added to soups, made into candies and creams, or ground into flour and used in breads, cakes, cookies, and even noodles.

Thinking of the chestnut as a nut will limit its potential as a food. Chestnuts are more of a grain than a nut. They have almost zero fat. Nutritionally they closely resemble brown rice. If you think of them as you do walnuts or pecans, you will only see them as a snack, just another tasty nut. But chestnuts are much more.

# Eating Chestnuts

Because of their high starch content, chestnuts have been a staple in the diets of people wherever they have grown.

They can be eaten and prepared so many different ways that I can only hope to illuminate a few of them here. The possibilities for chestnuts as a food are endless. In the kitchen chestnuts are as versatile as corn or flour.

## Harvesting

Harvesting chestnuts is one of my absolute favorite activities. Because the nuts are so beautiful, it's as if I can't help myself from picking them up. There is a deep urge to reach into a prickly burr and uncover the chestnuts inside.

When trees are not too big, I shake the branches to encourage as many nuts to fall as possible. This allows for prompt harvesting, which reduces the likelihood of animals and birds taking them and of weevils developing.

Some nuts will fall free of the burr and others will remain in it. Step on the burr and they should pop out. You can pick them up with your hands or a "nut wizard." This hand tool has a wire basket on the end of a handle and picks up any nuts it rolls over. It's pretty amazing that you can eliminate the need to bend over by using this simple tool.

There are also commercial mechanical chestnut harvesters. They operate similarly to machines that pick up golf balls. The truth is that hand harvesting is great fun, and it's not hard to find help. People will come from all directions to help harvest chestnuts if you put the word out. My friend visited a farm in Ohio that harvests 65,000 pounds of chestnuts a year—all by hand. It's just the farmer and two Amish families.

## Curing

When chestnuts first fall from the tree, they can be quite bland, and they need to be cured to sweeten up. If you store them in plastic bags in the fridge, curing takes a couple of weeks. As time goes on the nuts will get sweeter and sweeter—until late winter, when they start to sprout. If stored at room temperature, chestnuts will cure within a few days. There is no harm in eating chestnuts right off the tree, but their true flavor will not come out until after being cured.

## Storing

Keeping chestnuts from drying out is essential if they are to be roasted. They can be stored in plastic bags in the fridge. Check to make sure that excessive moisture does not build up in the bag, and vent the bag if necessary. They can get moldy in the fridge if it's not cold enough. The ideal temperature for storing chestnuts is 32 to 34°F (0–1°C).

Fresh chestnuts can also be packed in moist sawdust or sand in a root cellar at a warmer temperature. I have stored them for months at 40 to 50°F (4–10°C) by packing them into bins of sand. They keep well until spring when they sprout.

If chestnuts are to be dried or ground into flour, they can be hung in onion sacks and stored at room temperature for a long time. Dried chestnut kernels will keep well in a sealed jar for years.

## Roasting and Boiling

Roasted chestnuts are a great treat during fall and winter. There is no good reason why American city streets are not full of vendors pushing carts of hot roasted chestnuts. My family particularly enjoys eating roasted chestnuts in the woods, cooked on a bed of hot coals. At home we set them on top of the woodstove for about five minutes on each side. It is okay if the shells burn a little; the nut is usually fine inside. I prefer the shells to burn a little and get the kernels to be slightly overcooked. They taste really good when they are roasted to the point of being fully golden.

Recommended cooking times and temperatures vary, but there is no wrong way to roast a chestnut. Cook it less and it will be crunchier; longer and it'll be softer.

The shell needs to be sliced into, otherwise the nut can sometimes explode as steam inside tries to escape. Slicing the shell also makes peeling a lot easier. Many people slice an X-shape into the shell, but all that is necessary is one slice across the middle. Try not to cut through the whole nut, just the shell. A very sharp knife helps, and a chestnut knife

# Chestnut

Roasting chestnuts.

is very safe and useful. This knife is curved like a hook and is specifically made for cutting rounded nuts. Chestnut knives are inexpensive and readily available online.

You can also boil chestnuts. Make a slice across the shell, the same as you would for roasting. Boiled chestnuts are delicious and peel very easily.

## Drying and Shelling

Drying chestnuts allows them to be used in so many different ways, from soup thickeners, to stuffing, to flour. I have found that the easiest way to dry chestnuts is in the shell, by hanging them in onion sacks over the woodstove or laying them out on screens. It can take a few weeks of hanging for the nuts to be totally dried, at which point the shells will be very brittle. You will know they are dry when the nuts rattle in their shells when shaken. If you put them in a dehydrator, be very careful to keep the temperature as low as possible. Over 100°F (38°C) and they are likely to slightly cook and change color. Their flavor will be off if that happens.

Once the in-shell nuts are dry, I run them through a Davebilt nut-cracker. This is a small hand-powered device made for shelling hazelnuts and pecans, but it works well on dried chestnuts and acorns. A motor can be attached to the Davebilt. I attached a long handle to mine for increased leverage, and it is much easier to use than the short handle it comes with. You can also crack open the shells by simply crushing the dried nuts with a wooden stomper. If you place the dried nuts in a sack or pillowcase and then stomp, it really helps keep them in one place. Either way you will have to separate out the shells and pellicles (the skin between the nut and the shell) by winnowing or by picking out the nutmeats by hand. My friend John Walsh built an aspirator that runs with a shop vacuum. It's a simple setup that separates nuts and shells very well with minimal effort.

An alternative method is to peel them while they are fresh and then dry the nutmeats. To do this, cut the chestnuts completely in half and steam them for about one minute. While they are still hot, they will pop right out of the shell if you squeeze them with a pair of pliers or with your fingers. This method removes the pellicle easily and well. It only works while the nuts are still hot. Do small batches at a time to keep them hot. The peeled chestnuts can then be dried for later grinding, or

Dried chestnuts were run through an ordinary corn grinder. The flour was then sifted to separate the coarse meal (*top bowl*) from the fine flour (*bottom bowl*).

they can be ground up while still wet in a food processor. You can use this wet flour right away in baking or dry it for future use. You can also use these peeled nuts whole in a variety of dishes like stews and stuffings.

Personally, I like to keep a jar of dried, peeled kernels in the cupboard. Pour some in a pot to boil for 20 to 30 minutes and you have an easy, delicious snack.

## Flour and Meal

Making flour from chestnuts is the epitome of tree crops to me. Here is the actual grain falling from trees, the manna from the sky. This is not a new idea. Traveling to cultures outside the United States, we can find endless recipes calling for chestnut flour, especially in Italy and France. It is a very tasty, sweet flour that makes excellent desserts, but is also great for thickening stews, adding to stuffing, or just as a stand-alone bread. Chestnut flour is gluten-free. Breads baked with 100 percent chestnut flour resemble corn bread in texture. Cakes made with chestnut flour are dense and remind me of pound cake. It is a very high-quality flour with endless uses.

Chestnut flour even has an excellent shelf life. I have kept bags of it in the cupboard for up to two years without a hint of rancidity.

You can grind up chestnuts in an ordinary cornmeal grinder or a flour mill. To get them to fit into a flour mill, you will have to make the nuts smaller. I pre-grind them in a corn grinder. Some folks beat the dried kernels in a sack with a hammer. The dried kernels will shatter. If you only run them through a corn grinder, you will have a very coarse meal. If the resulting meal is then sifted, you will have a pile of fine flour and one of coarse meal. The coarse meal is an excellent soup thickener, is great in cookies, or can be boiled just like grits or oatmeal.

The fine flour is an excellent ingredient for so many baked goods. It does wonders for the consistency of cakes particularly. Chestnut flour is also great in all pastries, biscuits, and breads. It will not rise like wheat, so in many recipes chestnut flour replaces 50 percent of the wheat flour. However, I really prefer breads and desserts that are 100 percent chestnut flour. Chestnut flour is sweet and delicious by itself. There are endless chestnut flour recipes on the internet and in Italian and French cookbooks. *The Chestnut Cook Book* by Annie Bhagwandin is a great resource.

I am not a chef or a baker. I think every serious chestnut grower should be partnering with people who prepare food. Surely an entire culinary school could be devoted to working with chestnuts.

## Candied Nuts

There are several forms of candied chestnuts available today. The most famous are marrons glacés. Premium, specific nuts are selected for this process in which kernels are slow-cooked in sugar water until they fully absorb it and turn into the most beautiful-looking candy you've ever seen. Making marrons glacés is an art and a science. We've tried making them at home, but they've never come out as good as the pros' versions.

Chestnut cream is a product in which candied chestnuts are pureed into a cream. It's fantastic spread on toast and has the consistency of thick butter.

# Livestock and Wildlife

Another way to eat chestnuts is to eat the animals that eat them. Roughly 90 percent of our corn- and soy fields exist to feed livestock. This can be reduced in two ways. The first is by planting chestnut trees in and around pastures. The animals can harvest the mast with no work or processing on our part. Livestock can also be rotated through chestnut orchards after harvest to clean up any leftovers, thereby diminishing pest populations.

The second way to feed chestnuts to livestock is to harvest and dry the nuts. To make this commercially viable, mechanical harvesters and dryers are needed. This equipment already exists; it is just the orchards and hedges that need to be planted.

Commercial chestnut varieties are easy for pigs, sheep, and cows to eat, but they are too big for poultry. Allegheny chinquapins are much smaller and can be harvested by large poultry breeds without any processing.

For farmers looking to supplement grain feed, chestnuts offer a viable option. There is a long history of people feeding chestnuts to cattle, hogs, and sheep in Europe and the United States. Often animals were simply allowed to graze under trees, but nuts can also be harvested and dried for future feed. For larger animals, they don't even need to be shelled. For poultry, they can be ground up with shells included. The processing is quite minimal.

Chestnut trees attract and feed a lot of wildlife, including highly valued game animals like turkeys and deer. Every year hunters grow food plots of alfalfa, clover, and turnips to attract deer, but a few established chestnut trees would accomplish the same goal without the need for replanting. Chestnuts can be turned into meat, just like corn, alfalfa, and soy. They are a staple crop in every sense of the word.

## Comparing Chestnuts and Corn

Think of the ecology of a cornfield, especially a field of Roundup-ready corn that has virtually no weeds in it. There's really not a lot going on; not many places for hawks, butterflies, or honeybees. A chestnut orchard, on the other hand, is buzzing with life. There are 125 known species of Lepidoptera that feed on chestnut leaves.[4] The trees provide roosting and nesting sites for birds. In contrast with corn, they have extremely low fertility requirements. Chestnut trees can grow on steep hillsides where bare rock is exposed. They can heal and build soil instead of using it up.

I think it's pretty obvious to most people that the ecological diversity of a planting of chestnut trees is far beyond that of a field of corn. Of course, the big question people ask is, "Can an acre of chestnut yield as much as an acre of corn?"

It seems like a straightforward question that should have a straightforward answer, but it does not. An acre of fertile farmland is not equal to an acre of rocky hillside. There is also quite a bit of history to catch up on.

In 1929 when J. Russell Smith published his revolutionary work *Tree Crops*, corn and chestnut had similar yields per acre. He compared the yields of corn and wheat throughout the East Coast and Midwest with yields of chestnuts in Italy, France, and Corsica. Pound for pound, they were about the same. In some cases chestnuts were more and in some they were less, but overall there weren't any big differences. This was back in 1929, when corn and chestnuts averaged a little over 1,000 pounds per acre.[5]

Since then corn has been intensively bred to boost yields, with the full support of government agencies and universities. It is grown today with a full array of chemical herbicides and fertilizers. Its yields have become monstrous, to the detriment of the land it grows upon. This amount of breeding work and intensive farming has not happened to the chestnut. Despite the lack of breeding, chestnuts may still outyield corn today. A high-yielding cornfield now produces over 7,000 pounds per acre, while a high-yielding chestnut orchard yields 2,000 to 3,000 pounds per acre. So how can I say that chestnuts can still outproduce corn?

Corn requires fertile, well-drained bottomlands. It is grown in the richest fertile valleys that were formerly prairie. It is heavily grown in the Midwest upon some of the best farmland on the planet. Modern corn growing needs this; it needs flat places for large machines. It needs soils that can be cultivated again and again. Chestnuts, on the other hand, can grow anywhere. They can grow on the poorest of soils. Dry ridgelines, steep mountainsides, land that few people would deem worthy of farming is where the chestnut can thrive and produce heavy crops year after year with zero inputs. I would like to see a cornfield that thrived on a hillside without any weed control or fertilizer.

There are other layers to the productivity of a chestnut orchard. An acre of chestnut trees can easily accommodate grazing animals, shrubs, herbaceous plants, and mushroom logs underneath the canopy. Multiple crops can be harvested in the same space. An acre of chestnut trees can exist on landscapes where it would be virtually impossible to cultivate corn. So where is the acre of comparison between corn and chestnut? Is it in the very best fertile bottomland, or is it up in the hills with livestock occupying the same space? In many cases, chestnuts will crush corn in yield per acre. In all cases, they will have tremendously more biomass and biodiversity.

Because chestnuts can easily be grown in conjunction with other crops in the same space, and they can do so in places that corn could never grow, chestnut and corn yields are incomparable. They are very different uses of land, and the total amount of food produced will vary considerably depending on if you count all the crops grown inside a chestnut planting.

The technique of grazing animals beneath nut trees is not a new one. Today it is known as silvopasture; in ancient Portugal it was known as the dehesa system. When done well, by rotating livestock through

# Chestnut

the orchards, it is a proven sustainable method that has worked for thousands of years.

Annual grains are just that, annual. They die every year and have to be replanted every year. Chestnuts can live for thousands of years. They can start producing as early as age two, and will increase their nut production each year.

Farmers looking to convert fields of corn to chestnuts can do so with a technique called alley cropping. Here you plant rows of trees right into the field and continue to grow annual crops in between the rows. As the trees get bigger, the alleys of annuals get smaller, until eventually they are totally eliminated. This method can allow you to continue generating income during the transition period.

If we want to improve wildlife habitat, grow crops with ever-increasing yields, protect and build soils, and reverse climate change, then planting chestnut trees is one way to accomplish all of these wonderful and attainable goals.

# Cultivation

Cultivating chestnuts isn't difficult if a few things are tended to. The trees do best with abundant sunlight, good drainage, and protection from voles and deer.

## Pollination

Chestnuts are primarily wind-pollinated, monoecious trees (each tree produces male and female flowers). Though they are wind-pollinated, insects do play a large role in spreading pollen. They do not pollinate themselves, so two or more should be planted less than 50 feet apart. Occasional hybrid trees will have male-sterile

Chestnuts in bloom. They smell pretty weird, but they are beautiful and beneficial to many pollinators.

pollen. This means they can receive female pollen and make nuts, but they can't pollinate other trees. For this reason, I think it's a good idea for people to plant at least three trees together.

## Soil Preferences

Chestnut trees have a moderate-to-fast growth rate, similar to that of red maples. Of course, this depends on soil and tree health. They prefer good drainage and an acid pH, but this does not always have to be the case.

I have seen wild American chestnuts growing in low-lying swampy areas. At first glance, it seems that they can tolerate wet feet, but a closer look at the topography reveals something else. In older forests the ground is very uneven. As large trees topple over and are uprooted, a depression is created where the root mass once was. As the root ball breaks down, a mound is formed. This landscape feature is referred to as "pit and mound" or "pillows and cradles." It is associated with almost all old-growth forests.

Pit-and-mound landscapes are essentially covered with vernal pools and raised beds (see "Uneven Ground" in chapter 2). Chestnuts, like almost all trees, prefer to grow on the mounds. If you have a wet field and want to grow chestnuts—or really any other fruit or nut tree—then make some mounds. Because without decent drainage around their root crown, chestnuts will languish.

Virtually all literature on chestnut cultivation speaks of the trees' demands for acidic soil. However, I believe this depends on individual trees and species. I have on several occasions witnessed very healthy, mature chestnuts growing in alkaline soils. I believe that trees containing European and Japanese genetics are more likely to tolerate alkaline soil, but I have no scientific evidence to back this up. I also believe that high organic matter content in the soil will allow for greater flexibility of the tree to withstand differences in pH.

## Spacing

Conventional wisdom is to plant trees on 40-foot centers. This is a big area for a small seedling to take up. Eventually, of course, the trees will be very large. In some old European orchards, only four trees fit per acre. There are several ways people have chosen to use the space that is not yet taken up by the young tree.

One method is to interplant chestnuts with shorter-lived species such as peaches, raspberries, and asparagus. By the time the chestnuts are spreading their shade, these other plants will be on the decline. With the same ideology, some farmers use alley cropping to grow vegetables, grain, or hay in between the rows until the chestnuts get big.

My favorite method for spacing chestnuts is to breed them along the way. I plant rows of chestnuts with trees set out only a few feet apart. As the trees mature, I thin out all but the best-producing trees. This may sound expensive, but it is far cheaper than purchasing grafted trees, especially if you propagate your own seedlings. I keep rows 40 feet apart with trees set 2 feet apart within the row. These numbers are my personal preference; some growers go with more space, some with less. I heard Phil Rutter (co-founder of The American Chestnut Foundation) talk in an interview about planting chestnuts in a double row 18 inches apart. He described walking down the rows of trees and finding exceptional producers that stand out. I have not planted nearly as many trees as Phil Rutter, but I have definitely found trees with exceptional qualities by overplanting.

## Pests and Diseases

There are a few pests that can cause damage to chestnuts. A single grower is unlikely to encounter all of these. Chestnuts have far fewer pests to worry about than most fruit trees, but they are serious issues to pay attention to.

### Chestnut Weevils

The chestnut weevil is the most difficult and notorious to deal with, especially for the organic grower. There are two species, *Curculio sayi gyllenhal* and *C. caryatrepes boheman*. Essentially they cause the same damage and have the same life cycle.

When burrs are forming in late summer, the adult chestnut weevil will fly up and lay eggs into the nuts. The eggs won't do anything until the nuts fall from the tree. At this point they'll turn into a larva and start chewing their way through the nut, eventually emerging anywhere from 3 days to 6 months after nut fall. They look like disgusting small white worms and ruin the nuts. After they exit the nut, the weevils burrow down into the soil, morph into adults, and emerge the next year to fly up and lay eggs. Infestation in unmanaged plantings can reach close to 100 percent.

Two methods of organic control of chestnut weevil exist today. The first is sanitation. This means keeping a chestnut planting clean and free of fallen nuts. The life cycle of the weevil is broken if they cannot emerge from the nut into the ground. Every nut has to be harvested for this to work, and there cannot be unmanaged trees nearby (within a mile). Livestock can help thoroughly harvest all the nuts, but they must be picked up before larvae exit.

If orchard sanitation is not an option because of wild trees nearby, or because squirrels are carrying nuts away and burying them somewhere else, then hot-water treatment can be used. This also involves prompt harvesting of the nuts. When the chestnuts first hit the ground, the eggs are very small and unnoticeable. If you put the nuts into a hot-water bath the same day, then the eggs will be killed before they mature into the larvae. In order for the nuts to not be cooked or killed in the bath, the temperature should not exceed 120°F (49°C), and they should not be left in for more than 20 to 30 minutes.

The New York Tree Crops Alliance that I am a part of is currently experimenting with other organic controls. Unfortunately, almost no university research has gone into this. During the upcoming season we'll be spraying burrs during the month of August with a mix of clay, neem, and Grandevo. We will also be checking on the effects of spraying burrs with thyme oil. The weevil is a pest that must be kept in check for commercial production of chestnuts to succeed.

## Ambrosia Beetles

Ambrosia beetles tunnel into the trunks of young trees and form galleries where they cultivate pathogenic fungi. Untreated, ambrosia beetles will kill a young tree within a week or two. However, you can save trees that have been invaded. If you see small white sticks of frass perpendicular to the trunk, ambrosia beetles are in there. Immediately spray the trunk with pyrethrum—an organic product derived from chrysanthemum flowers. It is a broad-spectrum insecticide. Fortunately it has a very short shelf life in sunlight, only a few hours. Spraying it directly onto the trunk where the beetle frass is will take care of them and minimize harm to other insects. You cannot dig out the tiny beetles. Spraying is the only thing I have found that works. Ambrosia beetles only attack trees during a very short window, when the leaves are first emerging but not yet full-sized. Once leaves are full-sized, the trees are safe for the year. So checking young trees every spring during this time can save them.

Ambrosia beetles are the scariest pest for seedlings. They kill very quickly and can affect dozens of other species besides chestnut. There is a good chance you do not have them in your local area, but if you do, they need to be controlled vigilantly if you are to establish young trees.

## Asian Chestnut Gall Wasp

This wasp was accidentally introduced into the United States in 1974 on infested scionwood. The wasp lays eggs into the shoots of chestnut trees, and galls are formed. The larvae feeding inside the galls severely damage the new shoots of trees. Asian chestnut gall wasp is considered a major pest where it is prevalent, primarily in the Southeast. It can severely limit tree growth, and therefore nut production. Ozark chinquapin is the only species that carries any resistance.

Growers in areas without Asian chestnut gall wasp should avoid importing seedlings or cuttings from areas known to have it. This has been the primary way the wasp has spread.

## Chestnut Blight

*Cryphonectria parasitica* has been totally devastating to American chestnuts and Ozark chinquapin, and very damaging to European chestnuts. Chestnut blight is a fungus spread by wind, insects, birds, and nursery stock. It is native to Asia, where trees have evolved to live with chestnut blight. For trees that are susceptible to blight, the fungus feeds on the cambium layer and girdles the tree. Interestingly, chestnut blight cannot kill the roots of a tree. There are too many competing fungi in the soil for chestnut blight to live. Some people have used this understanding to keep American chestnut trees alive by applying mud packs to blight cankers. This allows the tree to heal over the canker.

There has been a lot of work done involving viruses that attack the fungus. This hypovirulence has proven an effective control in some populations for several decades. The Connecticut Agricultural

Experiment Station under the guidance of Sandra Anagnostakis cultures these viruses.

In my mind, the most practical and easiest method for dealing with blight is to plant resistant trees. There are endless sources for resistant hybrids and Asian chestnuts. There is a strong movement to find and breed resistant American trees. This is covered in greater detail in the section "Epic Saga of the American Chestnut."

Blight resistance is a spectrum. No trees are immune to the fungus. Most "resistant" trees will have some branches here and there die of blight.

If you live anywhere in the eastern US, there is blight in your area. Few people realize how common wild American chestnuts are. Also, the fungus is able to remain dormant for decades, and the spores can travel hundreds of miles. There are parts of the Midwest and pretty much all of the western US that are blight-free, but this will likely not last forever. Sooner or later someone will bring some blight over on a tree or a piece of wood or something.

## Ink Disease

*Phytophthera cinammomi* is just as terrible for chestnut trees as is the blight. Before the blight hit the United States, ink disease wiped out millions of trees in the coastal regions of the South. Ink disease has not been a problem in the North. Unlike chestnut blight, ink disease kills the entire tree, roots and all.

Ink disease thrives where soils become saturated for part of the year. There is no treatment. The best plan is to plant resistant trees in well-drained sites if ink disease is in your area.

All these pests and diseases may sound overwhelming, but it is most likely that a grower will only have to deal with one or two. It was 20 years before my friend Brian found chestnut weevils at Hemlock Grove Farm. I have never encountered Asian chestnut gall wasp or ink disease. Some folks don't have to worry about blight or ambrosia beetles. Whatever obstacles your trees have to overcome, it will be worth it in the long run. Chestnuts can live for 1,000 years or more and are bothered by far fewer pests than most fruit and nut trees.

# Propagation

Chestnuts are propagated either by seed or by grafting. While grafted chestnuts can have the benefit of bearing at a slightly younger age and bearing crops of large nuts, they have some serious disadvantages. The most important one is that they suffer from delayed graft failure. This occurs in about 50 percent of grafted trees. The top of the graft will die three to five years after it's been made because of incompatibility issues. Some growers state that this won't happen if the rootstocks are seedlings of the cultivar you are grafting. I don't think that's true, though.

Grafted chestnuts are also much more expensive than seedling trees. The main reason that I am not a fan of grafted chestnuts, however, is that they don't further the genetic expansion of the genus. If there is one tree that really could use our help in this, it is the chestnut.

Seedling chestnuts will not be exactly like their parents, but they are similar enough that it's not a big deal. Growing apples from seed is extremely variable, while with chestnuts you are going to wind up with a tasty nut just about every time. Sometimes the nuts will be smaller, but often they're going to be similar.

To grow chestnuts from seed, it's important to watch for three things: mold, early sprouting, and animals taking the nuts. Chestnuts need a cold stratification period, though some will sprout in the fall if left in warm conditions. To avoid early sprouting, I keep chestnuts cold. The fridge is barely cold enough. Outside is best (at least where I live). To keep rodents away from them, I place nuts in rodent-proof exclosures outside, in tote bins in the root cellar, or in buckets buried underground. In all cases I mix the nuts with damp sand (at about a 1:1 ratio of sand to nuts).

Sometimes they will sprout early. Be careful when handling sprouted chestnuts. If the radicles are long, you can trim them back. This will cause the root to branch out more. If the sprout breaks all the way off, they will make a new one, but this can weaken the tree.

If chestnuts get moldy, you can dip them in a mild bleach solution and repack them in fresh sand. Nuts that have been moldy but still sprout will grow rather bizarrely. They will often make a root that is no longer than half an inch and forms a swollen club shape. They are

Beds of chestnut seedlings at Twisted Tree Farm in early summer. By end of summer they will close the canopy. Once they go dormant in the fall, they will be dug up and ready for transplanting.

actually fine and healthy. If this happens, give them one more year and they will develop a large and normal root system.

When chestnuts sprout above the ground, watch out for marauding critters who will pull the tree up to get the nut. In some cases you will need to protect from rodents as well as jays and crows. I often raise seedlings under bird netting, with mousetraps all around. The first few years I raised chestnuts in my nursery, nothing bothered them. Now word has gotten out and all the jays, crows, chipmunks, and mice recognize chestnut sprouts. They can dig up trees at any point during the first year and occasionally into the second.

Some people raise trees in pots close to the house to keep a close eye on them. This is fine for just a few trees. Beyond that, you would need a good system for keeping critters away. Bird netting, mousetraps, and a distance from squirrel habitat work well for me. Growing chestnuts in the ground as opposed to in pots makes a dramatic difference. First-year trees in a well-tended bed will grow anywhere from 1 to 4 feet with a nice root system covered in mycelium. In pots, trees will get around 12 inches tall with a less exciting root.

You can also direct-seed chestnuts into their final planting spot. This can work very well. Be sure to mark where they are planted. I use a 1-foot-tall tree tube to mark the planting and protect the nuts. Trees establish with the best root system possible with this method and require the least amount of water as their taproots can extend down very far the first season.

## Commercial Possibilities

These are unlimited and very open right now. Around 95 percent of chestnuts consumed in the United States are imported. The demand for domestic nuts is nowhere near being met. I believe it is entirely reasonable that someone could make a living solely from growing chestnuts. Every grower I know sells all that they have. Thanksgiving and Christmas are peak times for fresh nuts. A cart of roasted chestnuts is almost guaranteed to do well in any urban center. I have heard of roasted chestnuts selling for as much as 50 cents a nut. Candied chestnuts, chestnut flour, cookies, and breads will appeal to a wide range of people including many Asians, Europeans, foodies, as well as gluten-free, paleo, and sustainably minded folks.

Chestnut orchards can be very productive, with yields ranging from 1,000 to 3,000 pounds per acre depending on site and cultivars. Wholesale, organic chestnuts can go for as much as $6 per pound; if you are selling retail, $10. And if you are processing them further, then the sky is the limit.

Chestnuts are one of the most dependable tree crops in the world. There is a chestnut industry taking off in Michigan, Missouri, Ohio, and soon New York. These states have been forming grower cooperatives that are helpful in getting nuts processed and marketed. Here in New York a group of us are just now forming the New York Tree Crops Alliance. I believe this is only the beginning; in the near future nut-growing cooperatives will cover the country.

When properly handled and prepared, chestnuts are delicious. Right now they are a high-priced specialty item, but that is only because more people are not yet growing and eating chestnuts in large quantities. We can actually accomplish amazing goals, like reversing climate change, improving wildlife habitat, protecting watersheds, and increasing biodiversity simply by eating more chestnuts. There is no good reason for our city streets, parks, yards, hedgerows, and farms to not be filled with chestnut trees. Every fall, kids and growers of all scales can be busy filling bins with the fruits of these generous trees.

CHAPTER EIGHT

# Apple
## The Magnetic Center

There is more to apple trees than you or I know. There are dozens of species around the world in the *Malus* genus. They can grow in swamps or rocky outcrops bearing fruit the size of a pea or a large potato. Skin color can be yellow, green, red, orange, and even blue. Both white flesh and red flesh are possibilities. Flavor can be horribly bitter, puckeringly astringent, candy-sweet, or complex and tart in ways supermarket consumers never know. But the apple tree is much more than its fruit. These trees are complex organisms with their own unique personalities and ecological roles.

Apples branch and branch and branch again. Their canopies are a tangled maze that reminds me of a human brain with all its tunnels

wrapped around one another. Lost under the shade of bigger trees, apples will become spindly, reaching for any light they can find. Out in the sun, they are spreading, vibrant beings welcoming sunlight, rain, wind, and the night sky.

The genus *Malus* is as diverse as the flavors of its fruits. It reminds me of the dog family. There is the wolf, from which domesticated dogs were bred. And so we have the wild apple, *Malus sieversii*, that is the parent to the domesticated apple. Many of these domesticated apples would be equated to dogs like the poodle and the shih tzu. Some are a bit more resilient, but none are like the wolf. And just as in the dog family, there are many wild offshoots of the apple. With dogs, there are foxes, coyotes, dingoes, dholes, jackals, hyenas, and several others. It is the same with apples. There are species originating from North America, Asia, and Europe. Some of these are hardy to $-50°F$; many make excellent disease-resistant small apples. Some are very rarely cultivated. You might not recognize them all as apples; some of them are only as big as a small blueberry, but they are indeed apples and they are very useful members of this incredible genus.

## The Wild Tree

Apples are generally small trees that sprout in old fields and hedgerows, and along roadsides. They are some of the toughest trees around, able to compete with the thickest of weeds and withstand endless browse. Wild apples tolerate some of the most extreme soil conditions, from overly wet to extremely dry, acidic to alkaline. They won't always thrive in these conditions, but they will often grow and produce fruit anyway.

Apple buds and leaves are highly palatable to deer. The twigs and bark are a favorite winter food for rabbits and many rodents. In the spring, when the flowers open, we can see dozens of species of insects gathering nectar and spreading pollen. On warm spring days in May when the apple trees are blossoming, I like to stand still in front of a big tree and let my vision go out of focus. After a moment, the activity is revealed. It is a mesmerizing event.

The leaves and fruit are also an important food for dozens of species of Lepidoptera and other insects.

## Apple

Like most trees, apples take a break after a heavy crop. They will generally flower very little the year after a large fruit-set. If there is a late frost that kills all the blossoms one year, the next year will see a very heavy bloom. Of course, a frost can wipe out a crop any year. Commercial orchards will prune branches and thin fruitlets soon after set to prevent trees from taking the next year off. They will also sometimes use frost machines to keep the blossoms alive by spraying them with fog or mist continuously on cold nights. In the wild, apples usually won't fruit every year, but as with all things apple, there are exceptions.

Wild apples are more common than many folks realize. There is a thicket of them on my property consisting of a few hundred trees. People often don't believe me when I explain that they are wild and not planted. In the spring I can see hundreds of wild apple seedlings in the grass as I walk across the fields. Wild apples are spread by deer and birds; they are one of the most successful trees to establish in old fields. They have no problem competing with such aggressive species as autumn olive, European buckthorn, multiflora rose, gray dogwood, and honeysuckle.

Apples can live for a long time; it's not uncommon to find trees over 100 years old around central New York State. I have heard of apples living for over 300 years in some parts of the world. When apples become old, they gather a lot of deadwood. Often whole trunks will be hollowed out. An apple tree can live like this for a very long time. It will send sprouts from everywhere and continue to grow even after it has completely fallen over and uprooted. These old large apple trees offer great denning sites for mammals and birds.

I have spent many hours perched up in a wild apple tree during the fall observing wildlife.

Wild apples growing at Twisted Tree Farm. This is what happens to many abandoned cow pastures in upstate New York.

It is truly a center of activity in the quiet of the woods and fields. Sitting completely still for hours in nature has become lost to most of our culture, preserved by a handful of hunters and a few Earth lovers. It is very good for the soul, though there are often stretches of boredom that are not easy to endure. Sitting in an apple tree when it is fruiting requires no such discipline. There will be plenty of wildlife for the short attention span if you are able to remain mostly still and hidden up in the canopy. Dozens of species of songbirds, chipmunks, squirrels, various rodents, skunks, possums, deer, and raccoons are all likely to stop by, and so are the predators of these creatures. Apples enliven the land wherever they dwell.

## The Domesticated Apple

Comparing wild apple trees with modern orchard systems is like comparing wolves with poodles. Domestic apple trees have been selected for centuries. Approaches to caring for them are as varied as the people who tend them.

### Conventionally Grown

The trend in commercial production is toward ever-increasing inputs and trees that cannot fend for themselves. In most orchards trees are fed, irrigated, and sprayed with multiple rounds of fungicides and insecticides, while all other plants are kept at bay with regular applications of herbicide. No grass grows under the narrow canopy of dwarf, trellised trees that hold up more weight in fruit than wood. Most conventional orchards comprise trees that literally cannot stand up on their own or even grow with a cover of sod. Toxic sprays are heavily used to fend off every insect and fungus. The work of the tree is just to make fruit. Everything else is taken care of by the orchardist. Mike Biltonen, an expert orchard consultant, describes conventional apples as being on a constant IV drip to keep them going.

### Organically Grown

The demand for perfect-looking fruit at a low price puts growers in a tough place. Many pests of apples don't affect the flavor of the fruit or the health of the tree, but they are unacceptable to consumers expecting a shiny apple.

# Apple

I have a T-shirt from my friends at Eve's Cidery. It has a diagram of an apple split in half with labels on all the parts of the fruit including the pericarp, calyx, radical, seed, and so on. The label on the skin of the apple reads UNREALISTIC EXPECTATIONS.

The East Coast is home to several significant pests that growers on the West Coast do not have to worry about. Almost all organic apples grown in the United States come from irrigated deserts in the West because of these pests. Trees out there are grown in semi-arid regions with no significant fungal issues and fewer insects. In the East growers have to be not only creative but also highly educated on the needs of the trees and the life cycles of pests.

East Coast organic apple growers are a subculture with such an in-depth knowledge of trees and pests that few people outside their circles can even understand what they're talking about. Their strategies are brilliant and creative. Originally, organic apple growers in the East were focused on how to kill pests with organic pesticides. They still are, but it has evolved into a practice in which tree immunity and orchard ecosystem health is supported to fight off the pests. These growers spray mixtures using clay, cultured microorganisms, plant extracts, and minerals. They plant mixtures of wildflowers and sometimes shrubs under the trees to encourage specific predatory insects and biodiversity. Their trees are the epitome of health, and they have to be to fend off some of these pests. But it is not just the vigor of the tree that is supported; their practices go far beyond that. Mating cycles of pests are confused with pheromones, and micronutrients are applied to boost immune activity in the tree at specific times of fungal and bacterial activity. Parasitic fungi are crowded out by beneficial fungi that are applied to leaf surfaces. These growers understand the life cycles of all the insects and fungi that feed on apple trees. Volumes could be written about their constantly evolving practices.

If you want to commercially grow apples that look like the ones you see in the store and you live in the East or Midwest, then it will take this kind of dedication. You can get away with a lot by having a site with good airflow and the latest disease-resistant varieties, but you will still need deep knowledge and understanding. Fortunately for the rest of us, there are many other reasons to grow apples than just for pristine-looking fruit on a supermarket shelf.

# Species

This list is by no means complete. It is here to give you an idea of some of the 50-plus species of apples found around the world.

## Sweet-Scented Crab aka Garland Crab (*Malus coronaria*)

This tree is native to the eastern half of the US. It is beloved for its unbelievably fragrant flowers. The fruits are small, about half an inch in diameter. They are yellow-green hard little things that are mouth-puckering. In the old days these apples found their way into ciders and jelly (they still can today). Garland crabs are small, gnarly trees that form thickets.

I think the most interesting thing about this tree is that it flowers much later than other apple trees. While domestic apples are in full bloom, the sweet-scented crab still looks dormant with completely closed buds. In a world of changing climate and wild fluctuations in spring temperatures, this seems a good tree to bring into the fold of breeding programs.

## Sargent Crab (*Malus sargentii*)

This amazing tree is native to Japan. It is widely planted for ornamental purposes. Sargent crab is extremely disease-resistant. The fruits are small—about half an inch in diameter—and fire-engine red. Fruits hang on the tree throughout the winter and are heavily fed on by birds. They're tart, but I enjoy eating them fresh. In the fall they are pretty mouth-puckering, but as winter goes on they mellow quite a bit. I have gathered and eaten countless tasty sargent apples standing in parking lots as late as March.

Sargent crab in late winter. These tiny apples have a sweet-tart flavor and are true living bird feeders.

# Apple

## Siberian Crab (*Malus baccata*)

Native to Asia from Siberia through the Himalayas, *M. baccata* reaches heights of close to 50 feet in the wild. This is the parent to the Ranetka apple that is widely used as a rootstock in far northern plantings, where extreme cold hardiness is a must. Siberian crabs are also extensively used in ornamental landscaping. They make tiny fruits and abundant blossoms.

## *Malus sieversii*

There is a place in Central Asia that is home to the largest genepool of apples in the world. In the Tian Shan mountains of eastern Kazakhstan, on the border of western China, stands the world's original apple forest made of *Malus sieversii*.

It is here that 300-year-old giants can be found bearing all types of apples, including immaculate fruit. Seeds from these trees were brought out along the Silk Road to Europe thousands of years ago. From there the apple has spread. But only a small portion of *Malus sieversii*'s genes were represented in this diaspora. The wild apple forests are large and remote. Most of what is there is unknown. Researchers started taking field trips in the 1990s into these forests and collecting seeds and cuttings from amazing trees. Trees that are resistant to many of the diseases orchardists spray to control.

A request made to the New York State Agricultural Experiment Station in Geneva, New York, will get you 100 free seeds sent from the Kazakhstan apples they planted. You can request the seeds here: www.ars.usda.gov/northeast-area/geneva-ny/plant-genetic-resources-research/docs/apple-grape-and-cherry-catalogs.

# Varieties

There are around 7,500 named apple varieties in the world. I'm going to list a few here to give you an idea of some of the diversity available. Understand that this is not even close to a comprehensive list. Also, understand that fruit will vary in quality from year to year and from location to location. Soils, climate, and even age play a role in flavor. For a longer list check out www.orangepippintrees.com, and for a really long list check out this seven-volume book that was recently published

Bramley's Seedling apples harvested at Eve's Cidery. Photo courtesy of Autumn Stoscheck.

just on apple varieties: *The Illustrated History of Apples in the United States and Canada* by Daniel Bussey.

- **Northern Spy:** An all-around great apple, good for fresh eating, cider, and cooking. Vigorous grower, ripens late season, large fruits somewhat disease-susceptible. Beautiful green, yellow, and red striping. This apple has been around since the 1840s for good reasons.
- **Liberty:** This apple was released from Geneva, New York, in 1978. It's highly resistant to all major apple diseases. Liberty has very good flavor; best used for fresh eating. Stores well all winter. Ripens late season.
- **Jonagold:** Delicious sweet dessert apple. One of the tastiest varieties. Generally susceptible to most apple diseases. Mid- to late-season ripening.
- **Dolgo Crab:** Good-eating crab apple; bright red 1-inch fruits make excellent sauces and jelly; good for fresh eating, too. Copious white blossoms, a great pollinator. Originating from Siberia, hardy to zone 2. Early ripening. Very disease-resistant.

## Apple

**Kingston Black:** Considered able to make a good hard cider by itself (typically, hard ciders need multiple types of apples in a blend). Originating in England, this apple is the standard hard cider tree there. Late-season ripening.

**Sweet 16:** Disease-resistant, sweet apples with yellow flesh. Mid- to late season.

**Black Oxford:** Purplish dark fruit. Great flavor, good for fresh eating, cooking, juice, and an excellent keeper. Mid- to late season.

**Bramley's Seedling:** Hands down the best cooking apple. Very large fruits put Granny Smiths to shame. Mid- to late season.

**Keepsake:** As the name implies, this is a very good keeper. Medium to small fruits; sweet flavor peaks during storage in midwinter; keeps until April. Very hardy. Late ripener.

**Hudson's Golden Gem:** Excellent flavor, russeted skin with yellow flesh; crisp and sweet with a nutty flavor. Late ripener.

**Pristine:** My favorite early apple, Pristine ripens in mid-July. Sweet and crisp, it stores well (especially for a summer apple). Very disease-resistant fruits, though leaves get cedar apple rust.

**Golden Russet:** Wonderful flavor, good disease resistance. Great for fresh eating, cooking, storing, and hard cider. Early- to midseason ripening. Like Kingston Black, this tree also can make a quality stand-alone hard cider.

**Roxbury Russet:** An old russet, possibly the first named variety of apple to be born in America. Good for fresh eating, cooking, hard cider, and storing. Mid- to late-season ripening. After hundreds of years, Roxbury is still highly resistant to scab, cedar apple rust, and fireblight.

**Akane:** Very disease-resistant, good dessert apple. Early- to midseason.

**Wickson Crab:** A great crab apple for fresh eating or for hard cider. Fruits are 2 inches, with wonderful color and flavor. Late season.

**Ashmead's Kernel:** An old variety from England with outstanding flavor that has a hint of pear. Stores well and is great for hard cider and fresh eating. Ripens late season. To many folks this is the best-tasting apple. Still fairly disease-resistant, even though it has been around for hundreds of years.

**Calville Blanc d'Hiver:** A 17th-century French heirloom. Makes lumpy green apples that are considered by many to be one of the best-tasting apples in the world. Slight citrus flavor, high in vitamin C. Ripens mid- to late season.

I could go on and on describing varieties, but then this book would never end. I have seen people so addicted to growing different varieties that they just kept collecting after they ran out of room. Each tree in their backyard orchard had a minimum of five varieties grafted on.

# Rootstocks

Cloned apples are the rule in commercial apple orchards. With known varieties, rows of trees can be harvested on a schedule. The qualities of the fruit are known and consistent from one tree to the next. This uniformity and predictability is also useful on the rootstock side of the tree. Rootstocks influence growth habit and flowering.

Dwarfing rootstocks are selected for weakness or poor compatibility with scionwood. They are the pampered poodles of the apple world. They are so weak that they often cannot live without irrigation and weed control. Some dwarfing rootstocks have a life span of less than 20 years. They cause the grafted top of the tree to flower at a very young age, around two or three years. The trees are easy to manage and train for orchardists. It's a lot easier to spray, pick, and a prune an 8-foot tree than a 30-footer. I understand why commercial orchardists use dwarfing rootstocks. They have tight margins and need high productivity at a young age.

Personally, I prefer real trees that can take care of themselves. I like apple trees that will be alive for my grandkids to climb and harvest wild honey from. I'd rather grow a tree that produces 1,000 pounds of fruit and acts as a magnet to wildlife and people for decades after I'm dead. There are clonal rootstocks that can do this, though seedling roots are more likely.

Clonal rootstocks have their origins in different agricultural experiment stations from around the world.

### Geneva

The Geneva series has been spearheaded by the work of Dr. Jim Cummins. Dr. Cummins has made endless crosses and experiments to create this series of rootstocks. In his work, beds of seedlings were sprayed with heavy doses of fireblight to discover resistant individuals.

Most of the Geneva series are dwarfs bred for super-productivity on high-input trellis systems. This is the modern wave in apple growing. Trees are planted a few feet apart in the rows and trained along a trellis. Irrigation and weed control are standard practice. What it looks like is walls of fruit. The per-acre yields on many of these Geneva roots are much higher than ever before seen. Fully productive crops begin by age three. However, many of these trees may have a life span of only 15 to 20 years, and most cannot stand up on their own without being staked. Many of the Geneva rootstocks have been showing up with unexpected problems in recent years, including poor graft unions and sudden apple decline. I will say that the folks in Geneva continue to develop and breed new rootstocks, so who knows what they will release next?

## Malling

The Malling series is an old time-tested group of rootstocks from the U.K. There are a wide range of dwarfs and semi-dwarfs. My top pick for strength and reliability is MM111. It grows to about 60 percent of the size of a seedling apple tree. It is a strong-rooted, tough tree. This is a rootstock that has proven its reliability. It is not a heavy producer.

## Budagovsky

The Budagovsky series also has a wide range of dwarfs and semi-dwarfs. The releases are not as old and tested as Malling, but they appear to be of good quality. Bud 118, developed in Russia, grows to 80 percent of a standard's size. It is very cold-hardy, tough, and precocious. Bud 118 encourages trees to flower around age three. It is less susceptible to burr knotting, as is frequently an issue with MM111. Bud 118 has red leaves, making root suckers easy to identify. What I really like about Bud 118 is that it combines the virtues of early bearing and strength. Generally, growers have to choose between the two.

## Antonovka

This strain of seedling apples originated in Russia. The fruits are used there for cooking and processing. In the US, Antonovka apples are used primarily as a rootstock or for wildlife plantings. They are extremely cold-hardy and adaptable to adverse soil conditions. Antonovkas are well-anchored, deep-rooted trees. As a rootstock, they are slow to influence fruiting, often taking 5 to 10 years.

Traditionally apple rootstocks were cloned in stool beds. They still are for the most part today, though tissue culture labs are starting to clone rootstocks. Almost all the clonal rootstocks in the US are grown in the Pacific Northwest. Millions of rooted layers are harvested off acres of clones. Fungicides, insecticides, herbicides, and chemical fertilizers are the rule.

Building your own clonal rootstock bed isn't hard, and it's a very productive use of space. Large commercial nurseries are harvesting 60,000 rooted stems per acre. That is a lot of trees per square foot. I set up a small stool bed for Bud 118 this last year. It was 3 feet wide by 15 feet long. I started with three individual two-year-old trees in the spring. By fall I harvested around 75 rooted stems, and this was only the first year. The root system of the stools will only grow stronger every year.

Stooling apple rootstocks is easier than most other trees. They root readily and freely into soil or sawdust piled against them. This is not true for all apple trees, though. One of the criteria in traditional apple rootstock breeding programs is ease of propagation. For a detailed explanation of stooling, see the propagation chapters of this book.

## Grafting Apples

Apples are one of the most forgiving trees to the propagator. They callus thickly and quickly under warm or cool conditions. Unions form easily. Bench grafting can happen anytime over the winter so long as the trees are dormant and you have a protected place to store them (I currently store mine in our unheated basement).

Top-working is done outside before apple blossoms open in the spring. Warm days when the sap is pushing are best—though I have seen fresh apple grafts survive temperatures in the teens.

I have found that it's easier for the tree to heal small wounds. So I prefer to graft onto branches or sprouts rather than onto the main trunk. After the grafts take on the smaller limbs, the rest of the trunk can be cut off.

Grafting apples is pretty easy compared with other species, and there are more varieties of apple available today than just about any other fruit.

## Seedling Apples

Search for "growing apples from seed" and you will see the same thing written at the beginning of just about every article: "If you grow an apple seed, it will make a tree that bears a different fruit than its parent." *This is true, and it's okay.* Diversity in nature is a wonderful thing. There are more shapes, tastes, and colors in the genetic pool of the *Malus* genus than we can even dream of. There are apples with blue skin and red flesh, some that ripen in early July, some that hang on the tree all winter tasting sweeter and sweeter. There are apples the size of a pea and others the size of a large potato. We can grow apples from seed and find wonderful trees.

Let's not forget that every amazing apple variety in this world started out as a seed.

For all the great advantages that cloned apple trees have, seedlings have quite a few themselves. First of all, there is no graft union. The entire tree is itself; any sprouts from the roots are true to the top. A tree girdled by rabbits in the winter does not need to be bridge-grafted; it can just grow back from the roots.

Seedlings can live a very long time. Apple seedlings have a life expectancy of over 100 years. Compare that with a cultivated apple on a dwarfing rootstock, which might live for only 15 years.

Lack of the graft union is essential to bigger size and longer life. Seedlings are also tougher. They can tolerate soils that would kill most cultivated rootstocks. On my farm, wherever it is wet and muddy we get wild apples springing up.

It is often said that 1 in 1,000 apple seeds will make a good apple. There is no way that can be true. I live in upstate New York. Wild apple trees are basically an invasive species here. I sample apples everywhere I go in the fall. I would say about one out of three trees is palatable. Maybe 1 in 20 tastes great. I just do not believe the notion that 1 in 1,000 trees is good. It appears to be just repeated information with no basis in reality. Taste lots of wild apples and you will find lots of great apples.

Also, virtually all apples can be processed into something edible and delicious—from dried fruit to applesauce to vinegar to hard cider to ethanol to pork and venison. All apples are suitable for at least one of these categories.

## Growing Seedlings

How does nature grow them? Very close together and crowded. If we are to plant seedling trees and expect 1 in 20 or 100 to be good, then we need to plant crowded. It is okay to plant 100 trees close together in a space where only 1 is wanted. The best way to do this from a management perspective in my mind is to plant a hedgerow. As time goes on the hedgerow is thinned out. The cut-back trees will regrow again and again unless they are mowed repeatedly in the summer. A few mowings in a year will kill most young trees.

You don't know what you'll find if you plant apples from seed—not just in the fruit, but also in the form. I hope to one day discover a timber-type apple and a timber pear. The wood would be highly prized by woodworkers, the trees grow fast, and they would be an excellent soft fruit mast for wildlife.

## Propagation by Seed

Growing apples from seed is not hard. The first time I did it, I just bit into an apple and took the seeds out. I planted the seeds in some potting soil in a window, and about a week later they sprouted.

Some apple seeds will sprout right away in warm soil, while others will require moist stratification. I use a couple of methods to stratify apple seed. With cleaned seeds, I pack them in damp sand in plastic bags in the fridge. With gathered fruit, I store seeds outside in the fall by letting the fruit rot outside all winter. By spring, the apples are mushy after all the freeze and thaw cycles. I then mix the rotten fruit into a slurry with a paint mixer. This slurry of pulp is raked into the soil about half an inch deep. Germination usually occurs within a couple of weeks.

I have also grown many seedlings from the pomace produced from cideries. I lay the pomace on top of a bed in the fall, chop it up with a hoe, and incorporate it into the top layer of soil. Seeds will sprout in the following spring.

# Apple

A nursery bed of apple seedlings at Twisted Tree Farm.

I find that storing the seeds in the fruit or in pomace is the best way to prevent them from sprouting too early. This early sprouting can be a problem with refrigerated seed.

Another way to grow apple seedlings is to look around large apple trees. You can often find seedlings growing on their own. Dig them up before the lawn mower gets them. They transplant easily and can sometimes be found by the hundreds under a single tree.

## Planting a Wildlife Plot

Apple trees are one of the best choices for wildlife plantings. Their fruit, buds, leaves, and blossoms attract a diverse array of creatures. Planting multiple varieties that ripen at different times offers so much more to wildlife than a block of the same clone. There are apples that ripen in July, ones that don't lose their fruit until April, and everything in between.

Many crab apples will hang on to their tiny fruits all winter. I used to think this was because the birds did not like them. I have since learned that those late-winter fruits are relished by songbirds returning from migration in the spring. A tree will hold on to this fruit through the endless onslaught of winter, dropping a few here and there occasionally. Then, in one quick swoop, a flock of cedar waxwings will harvest everything upon their return from warmer lands.

If you are planting trees for wildlife, then use seedling rootstocks for long life. Use a multitude of varieties with different ripening times. Crab apples in both wild and domestic form are often the true champions in productivity for wildlife.

## Wood

The wood of the apple tree is another of its gifts. It has a warm rich color and smell. The density is very high, coming close to hickory. Applewood burns very hot and long. It makes an outstanding wood for carving and is somewhat durable. I have seen apple trunks and branches with perfectly sound wood after being outside for several years.

Apples grow fast and can endure repeated and hard cutting. They coppice well and can grow back from large wounds easily. They have a powerful ability to heal themselves. The branching can be so tremendous on apples that it's difficult to harvest large amounts of wood. I have often daydreamed of finding an apple tree with a form like an ash: a tree that grew straight up, making a thick trunk; a log of apple. Such a tree could grow 70 to 80 feet and compete well in the hardwood forest. Every two or three years it would rain down a crop to feed the animals. Floors, tables, or cabinets of applewood would be just as striking as those of cherry and mahogany. I believe this tree exists and that a forest of mixed hardwoods with apple is a true possibility. One inspired person could create a viable population of timber-type apples. They would grow out thousands of seedlings every year, each generation becoming better and better. Eventually walking through a forest of oak, hickory, chestnut, and apple would be a reality.

## Using Apples

Certainly an entire book could be written on all the uses for these magnanimous beings. The apple is much more than a snack in a kid's lunch box. Hard cider, juice, vinegar, ethanol, sauce, butter, chips, and an endless array of desserts are just some of the uses for this fruit that is revered by our civilization.

Apples store well in cool conditions. A normal root cellar is too warm for them, and even good keeping varieties will become mealy after a month. Storing apples at 32 to 34°F (0–1°C) will keep certain varieties in perfect shape for six months or more.

Claude Jolicoeur recently published *The New Cider Maker's Handbook* for folks wanting to delve deeply into this subject. Hard cider is an old tradition that is finding new life today. An industry is forming across the US, and old knowledge is being regained and expanded upon. There are many types of hard cider and they are not all the same. They are as varied as beer or wine in flavors and quality. Unfortunately, many of the more visible hard cider brands in stores are made by large companies using sweet apple juice concentrate from China. If you are interested in hard cider, I recommend you try traditional cider made by a small cidery blending bitter apples high in tannin and acidity. Just as wine from Concord grapes is not the same as wine from true wine grapes, so it is with apples. There are trees that produce fruit with the best qualities for fermentation. Real hard cider doesn't taste like Mott's apple juice.

Autumn Stoscheck from Eve's Cidery gathering crab apples. Autumn works along with her husband Ezra and a small crew to produce one of the highest-quality ciders in the world. Photo courtesy Ezra Sherman.

Applesauce and dried apples are what I have done with most apples I gather. These are pretty simple foods that many people are familiar with, and it's easy to find information on these products. Here are a few tips that you may find useful.

With sauce, I don't peel the fruit. I cook the apples down with the skins on them; sometimes I remove cores and sometimes I just throw the whole fruit in the pot. Either way, I let the fruit cook on low heat in a stockpot with a tiny bit of water in the bottom. When the fruit is soft enough that it mushes easily, then it's done cooking. While it's hot, I run it through a food mill to remove seeds and skins. Applesauce cooked with the skins on this way often comes out pink as opposed to the pale yellow we see in stores. I have never found a need to sweeten applesauce made this way, and I do love sugar.

Drying apples is easy, and you don't have to dry them in ring form. I just slice apples into pieces and lay them on screens either in a dehydrator or above the woodstove. The key is to not slice them too thick or too thin. If they're too thick, they won't dry well. If they're too thin, they'll be brittle, paper-like chips that aren't much fun to eat. Experiment with the thickness of the slices and you'll find a size that is perfect—a little chewy and not too brittle. Making your own dried fruit is very satisfying and makes economic sense. There have been years that I've had several gallons of dried apples in the cupboard; they make an easy snack for the family. It would be cost-prohibitive to let everyone freely eat them if we purchased them in the store for $20 a pound, however.

You can also dry applesauce or apple butter. Just spread it out on a cookie sheet suspended over the woodstove or placed in a dehydrator. Homemade apple fruit leather is awesome.

# Pests and Diseases

Here are some of the most basic and common pests to manage if you are looking to just keep your trees healthy enough to produce fruit for processing or wildlife. One thing to keep in mind is that not all disease-resistant varieties are insect-resistant.

The telltale sign of an apple borer: orange frass.

### Round-Headed Apple Borer

If you only pay attention to one pest, this is the one to focus on. Apple borers kill young trees quickly. Eggs are laid throughout the summer months in the bark. The eggs turn into larvae, which tunnel into the tree. Damage can be severe enough that the tree just breaks off at the base from all the tunneling. Check the trunk for orange frass (insect poop). It is usually low on the trunk, or sometimes an inch below the soil. Dig out the grub with a pocketknife. You have to find it and kill it. Otherwise it will keep chewing through the tree. It will seem like you are hurting the tree by digging into it with a knife, but if you don't get the borer, it will often kill the whole tree. No damage you do with the knife will be worse than leaving the grub(s) alive.

## Apple Scab

This is a fungus that covers leaves and fruit with dark lesions. Damage can be severe enough to block most photosynthesis and render fruit inedible. Spores overwinter in fallen fruit and leaves. One management strategy is to rake up or mulch over all the leaves in fall. Another is to plant scab-resistant varieties. Understand that apple scab, like most fungi, is evolving rapidly. A tree that has full resistance now may or may not be resistant in a few decades. As resistant trees show up, the fungus can mutate. This ongoing drama reminds me of the relationship between deer and wolves. Each species makes the other better by improving itself.

## Fireblight

This very serious bacterial disease can kill branches or entire trees very fast. Affected trees look as if they have been burned: The ends of their branches are black and curled into a shepherd's crook. Fireblight can be controlled by planting resistant varieties and pruning out and burning all infected wood.

## Cedar Apple Rust

This rust-colored fungus can cover leaves and fruit, causing extreme damage in some cases. It has a complex life cycle in which it uses two hosts—cedars and apples. However, it is very important to understand that the word *cedar* has been misused and often causes confusion. True cedars are those in the genus *Cedrus*; they are from the Mideast (as in, the cedars of Lebanon). The name *cedar* in America generally refers to the genus *Thuja*, sometimes known as arborvitae. But it is the red cedars in the *Juniperus* genus that actually host cedar apple rust. This is important to know if you are planning on removing host trees in the area.

It is also important to know that not all red cedars (junipers) are susceptible to the rust. Many are resistant. Junipers are commonly planted in landscapes and abound in the wild. Many growers recommend removing all "cedars" in the area. However, blight spores can travel for miles, though infestation is more severe where the trees are close together. If you do want to cut out red cedars or junipers, it would be a good idea to be sure you have the right tree and watch for galls. Red cedar trees that are hosting cedar apple rust will be covered in bizarre-looking orange galls.

The best way to deal with cedar apple rust is by planting resistant varieties, of which there are plenty to choose from.

## Commercial Possibilities

Apples offer many options for someone who wants to make a living from trees.

### Nursery Stock

Certainly there is a demand for seeds, cuttings, rootstocks, and trees. Apple trees are planted regularly by orchardists, homeowners, ornamental gardeners, homesteaders, preppers, wildlife proponents, and municipalities. Apples are one of the easiest trees from which to sell plant material. They are a large part of my nursery's income.

### Hard Cider

This is a very fast-growing industry right now. Cideries are springing up all over the Northeast and in many other regions. You can start your own or sell fruit to one if they are in your area. Several cideries near me buy wild fruit. They want fruit that has a high level of astringency or bitterness. Make sure you know a fruit is wanted before you harvest. Not all wild apples are preferred by the cider makers, but many are. Sixty cents a pound is what we've been paid. It's hard work, but totally possible to make $1,000 a day. Laying down tarps and shaking big old apple trees is a lot better than going to a job you don't believe in.

### Sauce/Dried/Leather/Juice

Though I admire the work of organic orchardists, I have never been interested in trying to grow apples for sale as they are. Processing will render all blemishes invisible. Processing will also allow you to use wild

## Apple

The cider barn at Eve's Cidery. There are gigantic corporate producers of hard cider, but cider making is also a burgeoning cottage industry.
Photo courtesy Autumn Stoscheck.

apples that may have a more astringent flavor. Sometimes simply drying or cooking fruit will remove all trace of astringency. If it doesn't, honey works well. The price for processed fruit is very high and always will be. You do need access to a commercial kitchen, but you don't have to build one to start out. You can rent space in many commercial kitchens for an hour at a time. Often, they are just in people's houses. I believe processing apples into sauce or chips or juice is the best way to make a living from an apple orchard. You can cut your costs way down by eliminating sprays, reducing pruning, and harvesting any fruit that's halfway decent. You can harvest fruit by shaking it down onto tarps.

You can compete with larger operations by reducing your costs. Fruit can be harvested from wild trees that cost you nothing to plant or maintain. You can even collect fruit from people's yards. It's amazing how many millions of pounds of apples fall to the ground uncollected in the suburbs of America every year. Your biggest costs would be the initial purchasing of equipment for juicing, fruit pulping, or drying.

If you have been inspired to grow apples but feel intimidated by the pest issues they face, then I recommend taking a close look at processing fruit.

## Fresh Fruit

This is an enormous industry, and there is an insatiable demand for good apples. If you choose to go this route, you should talk with other orchardists (as many as you can). You should also read Michael Phillips's books: *The Apple Grower*, *The Holistic Orchard*, and *Mycorrhizal Planet*.

Apple trees bursting out of hedgerows with their blossoms in spring, bent to the ground with heavy fruit in the fall, the smell of cider in the barn. How missed our apple trees would be if they were to disappear tomorrow! These trees are a gift to the world; they are magnetic epicenters of life. Everyone who plants an apple tree can know that they have left something wonderful behind (so long as it's not on a dwarf rootstock!).

CHAPTER NINE

# Poplar
## The Homemaker

P oplars are some of the fastest-growing trees in the world. They can tolerate the worst conditions and are heavily favored by wildlife. In places where land has been degraded or is falling apart, the poplars can rebuild. They produce tremendous amounts of biomass, feed unbelievable numbers of insects, birds, and mammals, and suck tons of carbon out of the sky like gigantic outstretched vacuums.

## Species

There are 35 species of poplars in the temperate world. These are the ones I am most familiar with.

### Quaking Aspen (*Populus tremuloides*)

Also known as the trembling aspen, quaking aspens grow across a wide range. They are native from Nova Scotia across most of Canada to Alaska, extending south through Pennsylvania in the East and down through the Rockies out west.

Their name comes from the way the leaves move in the wind. Like all poplars, quaking aspens have a flat stem. They flutter back and forth very fast in any breeze. The sound in a grove of aspens on a windy day has got to be one of the best sounds on Earth.

Quaking aspens have a very white bark that is often confused with birch bark. Birch trees have lenticels (horizontal lines) on their bark, however; aspens do not. Quaking aspens also have a white powder on their bark (some trees more than others; western trees seem to have more of this powder). It can be wiped off with a thumb; it feels a bit like

Quaking aspen grove in upstate New York. Notice the thick undergrowth of shrubs as well as the snags with woodpecker holes.

chalk powder. Birch bark, on the other hand, is full of lenticels and peels off in strips with no powder.

The trees grow fast, but not very tall, around 40 to 50 feet. Quaking aspen is the most prolific root-suckering tree I know of. Aspens send out horizontal roots that grow like mint, but on the scale of trees. They can form large colonies through their root systems. The most famous of these is in Aspen, Colorado, which is home to a 3,000-acre grove. This grove is one organism—a single tree with a massive root system and many trunks. If you don't think nature creates clonal monocultures, the aspen may cause you to question that.

Aspen groves can live for a long time. There is one in Utah that has an 80,000-year-old root system. None of the trunks are very old, but the roots continue to send new shoots up.

## Quaking Aspen Habitat

These trees create different ecological conditions depending on if they are growing in the western mountains or the eastern forests. They both require abundant sunlight. Out west they form large pure stands and persist for very long periods of time. You can see huge groves of them in the Rockies in the fall, when their yellow color really stands out. In the East it's a completely different story. Here quaking aspens appear in old fields and disturbed areas. They have only a certain amount of time before bigger hardwoods take over. Quaking aspen creates an outstanding situation for hardwoods to grow.

They will spread into the worst spots—wet, muddy fields, piles of sand—anywhere there is sun. They grow fast and create a lot of biomass. Their leaves and wood decompose rapidly. Their canopies cast a light shade. In the middle of an aspen grove, you will find grasses and shrubs and tree seedlings. This light shade reduces stress for many species that would have a tough time establishing themselves in an open field. After around 50 years, the bigger timber trees will have taken over and the aspens will have died out, their trunks falling over as a massive carbon mulch for the soil. They set the stage for fields to turn into forests. Aspens will persist on the East Coast if there is continued disturbance, but without it they'll give way to the giants.

In any situation, quaking aspens create some of the best wildlife habitat in the temperate world. Their buds, leaves, bark, and catkins are highly palatable to many species of birds and mammals. While the groves are young or after a large disturbance is the time when groves offer the most. When aspens are cut, burned, or disturbed, they will send up a huge amount of shoots. These can grow very densely, offering great cover and a lot of food at the same time. The buds and catkins are a big winter/spring food, highly sought after by wildlife of many types. Grouse, deer, and beaver are probably the most notable species to thrive off aspen groves, but there are also countless songbirds, elk, moose, porcupines, rabbits, and hares. There are virtually no herbivores that do not feed on poplars, and quaking aspen offers the most food of all the poplars. It grows the most stems with its endless suckering habit.

Quaking aspens are so tough, they can be cut ruthlessly and repeatedly and show no loss of vigor. I have never raised livestock, but if I did I would have a hedge of aspen somewhere for them. The trees can withstand repeated browse. I have seen groves of aspen establish themselves in areas with absurdly high deer populations.

In a world of degraded soils and dwindling numbers of birds and amphibians, the aspen offers us a low-maintenance approach to healing damaged lands.

### Managing Quaking Aspens for Wildlife

Older aspen trees offer the best den sites and woodpecker food. Younger stems offer better cover and easier-to-reach buds, catkins, and leaves. Grouse do especially well in these thickets of young aspen stems. Younger stems also have more tender bark for mice, voles, beavers, and rabbits. It is best if you can have a mix of old stems and young ones.

Young sprouts will not grow under the canopy of older trees. There needs to be an open side that the grove can spread into. You can maintain this open area by cutting it down every few years. After aspens are cut, they will send up a lot of stems the following year. It's better to cut trees while they are dormant and their reserves are in their root systems. In some parts of the country, aspens are managed with fire to make them productive. Productivity is measured in stems per acre.

If beavers are present in a grove, they will cut every tree within reach. For trees that are farther inland, they will dig canals and get all the aspens they can. The trees respond with abundant sucker shoots. These grow on a rotating basis as the beavers repeatedly harvest them.

I believe quaking aspen creates some of the best wildlife habitat of any plant, offering cover and year-round food. At the same time, aspen groves are some of the most soothing places to walk through as the filtered sunlight and rushing of the wind through their shaking leaves make their way to your senses.

### Eastern Cottonwood (*Populus deltoides*)

These are the giants of the poplars. Eastern cottonwoods can grow to over 100 feet tall. They grow rapidly—enormous trees are often no more than a few decades old. A friend of mine once cut a cottonwood growing too close to his house. The tree was close to 20 feet tall and had a trunk the diameter of an average adult man's leg. When we counted the rings, it was hard not to laugh. There were only four. They were spaced over an inch apart. This tree can put on wood at a staggering rate.

Cottonwoods can spread clonally like the quaking aspen, but not nearly as prolifically. In general they spread by seed, releasing it on warm sunny days in late spring. The cotton-like fluff looks like snow as it spreads everywhere. Seeds that land in waterways are carried off until they are washed ashore on some gravel bar or bank. This is how they are most successful at propagating themselves.

Cottonwoods drink copious amounts of water. They are generally found along riverbanks, canals, and abandoned gravel pits. They drink so much water that they can dry wet spots up. Once while working in the tree service, my colleagues and I were taking down a large cot-

I have always been struck by the beautiful shapes of eastern cottonwood.

tonwood in the summer. My co-worker's chain saw touched the trunk of this giant—and out came a fountain of water. It looked like a drinking fountain, only about twice as thick a stream of water. It poured out in this fashion for several minutes. I felt compelled to taste it and almost puked when I did.

Trunks over 4 feet in diameter are not rare. These giants only live to be around 70 to 100 years old. As they decline, they leave a wake of biomass behind. Large limbs and trunks become riddled with woodpecker holes as the birds easily tear apart the soft wood to find grubs inside. Denning sites for mammals are common in old cottonwood trunks. The wood breaks down pretty fast, and is an outstanding builder of soil.

The trees coppice well. They can withstand high browse pressure. Their leaves and buds are just as popular among wildlife as the aspens are.

Cottonwoods are a supreme biomass tree. They will feed and house an abundance of wildlife while enriching the soil and drinking excess water in wet areas.

There are several other species of cottonwood ranging across North America. Most grow along waterways. In some arid regions out west, they will dominate riverbanks.

## Bigtooth Aspen (*Populus grandidentata*)

These trees grow sporadically throughout the eastern woodlands in forests of various ages. I have found them in both young and old woodlots. They are usually found as just a few scattered specimens, but in forests that have seen a lot of logging they are often more common. They provide the same food and habitat as the quaking aspens and cottonwoods, but on a smaller scale since they don't grow in dense groves.

Bigtooth aspen has possibly the most beautifully colored leaves in fall. Colors can range from iridescent shades of purple and pink to yellow and red, all on the same leaf. The margins of the leaf are covered in big teeth (no kidding).

I thought of bigtooth aspens in the past as low-value trees in the woods, but I have come to see that, scattered throughout the forest, they add an element of diversity. The trees feed an abundance of Lepidoptera and other insects. Birds and mammals use their trunks, leaves, catkins, and buds.

## White Poplar (*Populus alba*)

This tree is native to Eurasia. It was once widely planted as a windbreak and for ornamental reasons. Due to its naturalizing habit, it is rarely planted today; it's now viewed as a noxious weed. The leaves of white poplar are silver on the underside and very beautiful in the wind. Non-native poplars do not feed native Lepidoptera and other insects the way native trees do.

## Hybrid Poplars

These have been bred by crossing European poplar (*Populus nigra*) with cottonwoods. The trees grow extremely fast, often reaching 30 feet in just three or four years. The wood they produce is harder than the native poplars and is suitable as a fuel wood. An acre of hybrid poplar will produce more BTUs than an acre of oak forest[1] (though with far fewer ecological benefits).

Hybrid poplars grow so fast that they are typically harvested on a seven-year rotational basis. Logs are used for making plywood, biofuels, and paper (yes, at age seven!).

Plantations of hybrid poplar are always made of clones. Some of these large plantations out in the Northwest are thousands of acres and support little wildlife. They are voids for biodiversity. This is not the fault of the tree, but a mistake of management. The difference between clonal blocks of native poplars and hybrid poplars is in the origin of the trees. Native insects have evolved to eat native poplars. There are countless species that will only feed on poplar leaves. These insects are the backbone of a food web that supports birds. Hybrid poplars do not feed insects the way aspens do. In fact, many newer hybrid poplar clones have been genetically engineered to avoid being food for Lepidoptera (butterflies and moths). These trees have had Bt (*Bacillus thuringiensis*) added to their DNA. Bt is a bacterium that is regularly used to kill Lepidoptera. The result is a silent forest. Transgenic poplars are not available to the public in the United States at this time. They are widely planted in afforestation projects in China.

Hybrid poplars are more suitable in mixed plantings or as hedgerows and windbreaks on diverse farms. They make fast-growing hedges, offer a massive amount of wood at a young age, and can be cut again and again. For details on growing and using hybrid poplars, I recommend Peter Greatbatch's *Practical Guide to Renewable Energy Using Hybridized Hardwoods*.

## Propagation

Quaking aspens grow easily from root suckers dug up from the edge of a grove.

Cottonwoods and hybrid poplars root easily from hardwood cuttings. They require no treatment; just stick them in the ground as far as you can. They root so well that truncheon cuttings can be used. These are large cuttings that can be as thick as a small log and 8 feet long.

Cottonwoods have been used on erosion control projects. Live stakes are hammered into the ground. Then smaller-diameter branches are woven through the large stakes to create a living wall.

I have never grown poplars from seed, but they regularly volunteer themselves in my nursery beds. Seeds require no treatment and will sprout very soon after landing on the surface of moist exposed soil. The seeds have almost no shelf life and should be used fresh.

## Commercial Possibilities

On an industrial scale poplars are widely used for paper, biofuels, chipboards, boxes, matchsticks, paneling, furniture, and a host of other

Bigtooth aspen logs stacked up in the state forest. The sheer amount of regenerative biomass that poplars offer is staggering.

products. To the homesteader they can offer a profitable side business. Because they coppice so well, all the poplar species can be used for any project involving biomass. The stems are not thorny or heavily branched, making them easy to handle. A grove of aspens, cottonwoods, or hybrid poplars can be the foundation of a biochar business. The whole grove can be cut every year, or sections of the grove can be rotated through over a few years to allow maximum wildlife habitat. Poplar stems are a joy to handle if you are used to moving brush.

Plant material such as cottonwood and hybrid poplar cuttings, aspen seedlings, and rooted plants are used widely in conservation and timber plantings. Restoration projects are a big business—there is a large market for poplar plant material.

CHAPTER TEN

# Ash
## Maker of Wood

Ash trees are the quiet workers of the land around here. They are not flashy or appreciated by many people. They just quietly grow in every vacant nook and cranny. Ash trees tolerate the worst conditions without a complaint. They are simple-looking trees with a profound silent power. Fast growth, excellent multipurpose wood, wildlife food and habitat, and adaptability are the subtle ways in which ash trees contribute to the world.

When I was first learning to identify trees, my old friend Luke helped me a lot. I remember one time he said, "See how that tree has thick twigs that are opposite? That's ash." I thought to myself, *That's easy*. A few days later I felt pretty frustrated. I was seeing trees with thick opposite twigs everywhere. There was no way they could all be ash, I thought. Well, they were. Ash is ubiquitous. It grows in virtually all locations and soil types in the Northeast. It is one tough tree that can grow very well in saturated soils, dry ridges, shady groves, competitive fields, sidewalk cracks, and even in the deep woods. Ash trees have very tough, aggressive root systems. I have dug many ash trees out of gardens and am always impressed by the size and strength of the roots.

Ash trees can reach magnificent sizes, even in poor soil. To any wood cutter, they are a gift: simple structures with few branches that are easy to handle. They cut and split like nothing else as their grain is often arrow-straight.

Ashes are one of the fastest-growing trees out in the open sun, but they can tolerate significant shade, too. In the understory, they will grow very slowly, waiting for the canopy to open. Seedlings can persist for decades under very deep shade. Once I was helping a group of kids make longbows. We were thinning out a grove of what looked like very young ash growing extremely close together. One of the kids and I picked a tree that was pretty straight and about the diameter of a baseball bat. After we cut it down, I counted the rings. It was 40 years old! Another time I was with a mentor, Ricardo Sierra of Hawk Circle, who was cutting down some larger ash trees for a project. The trees were about 40 feet tall and as big around as a large man's thigh. After the trees came down we counted the rings. They were all seven years old. That is the difference between an ash growing in sun and one biding its time in the shade.

I've always appreciated ash trees for their firewood, but since I have witnessed the beginning of their possible demise, I see that there is more to them than just that.

# Species

There are 45 to 65 species of ash found around the world, growing in such diverse places as Maine, Mexico, Southern Europe, and California.

### White Ash (*Fraxinus americana*)
Native to the eastern half of the United States, this is the tree I know best, since it is so common in all directions from my house. It is found in the valleys, along the roads, up on the ridges, and even on the edge of swamps. It is the biggest and most common ash in the Northeast. White ash is also the most adaptable of the different species around here. It is an excellent pioneer species, able to colonize fields at an astonishing rate.

### Green Ash (*Fraxinus pennsylvanica*)
Native to the eastern half of the US, green ash is always smaller than mature white ash in my experience. It is less common and grows mostly in very wet fields or swamps. Its wood is not as dense as white ash, but it is still decent enough for lumber and firewood. Green ash was commonly planted as a street tree before the introduction of emerald ash borer. It has the ability to grow well in anaerobic soils and can tolerate the salt and compaction that happens in urban plantings.

## Ash

Green ash growing in a swamp along the edge of Seneca Lake, New York. This species can tolerate extreme drought as well as prolonged flooding.

### Black Ash (*Fraxinus nigra*)

Native to the northeastern US, black ash grows primarily in swamps. Its bark is quite different from any other ash. It is smooth at the root flare and then furrows a foot or two up the trunk. The bark is corky, like American elm. It feels like Styrofoam when you push on it. Black ash has a long tradition in basketmaking. Logs are cut, soaked, and pounded with a sledgehammer. After a serious amount of pounding, strips of wood begin to separate along growth rings. These strips are woven into some of the strongest baskets in the world. Black ash pack baskets are of an unmatched quality and durability.

## Pests and Diseases

It is tough to learn that such an unassuming, tough, quiet tree should appear so doomed. I don't know if there will be many ash trees around here in the future. I have taken them for granted for so long: one of

Typical emerald ash borer sign. This tree's outer bark is being stripped by woodpeckers going after the borers.

the most useful and boring trees that is everywhere. Now they may be gone in my lifetime, and in a couple of generations few people will miss them. Of course it is impossible to see the future, but it looks like a tough road ahead for the ash.

The biggest threat facing ash trees is the emerald ash borer (EAB). This bug was brought from Asia by accident in wood pallets or packing material in 2002. It was first discovered in North America in Michigan, where it quickly wiped out 80 million trees. EAB is currently in 31 states and has killed hundreds of millions of trees. Research is being done on introducing a parasitoid wasp that is EAB's natural predator in China.[1]

Ash trees resistant to EAB have been discovered in Ohio and Michigan. It appears that about 0.5 percent of ash in some populations are unaffected by the borer. These trees are a genetic treasure. Foresters and regular people have been collecting and saving seeds. Surviving ash in areas that have been decimated by EAB are known as lingering ash. They are often isolated from one another and cannot repopulate without human intervention. Aside from pollination, such a large decrease in population can lead to a genetic bottleneck. Now is the time to be identifying lingering ash and growing seeds out.

When chestnut blight swept through the East Coast a century ago, foresters recommended cutting all the trees before the timber was ruined. Today the same is being recommended for ash. We will never know what genetics were destroyed during the collapse of the American chestnut. However, it is not too late to avoid the same mistake with our ash trees. The US Forest Service and MaMA (Monitoring and

Managing Ash) are two organizations that are working to collect seeds and identify resistant individuals.[2]

Though EAB is an enormous threat to ash trees, they have already been dying in large numbers because of ash blight, aka ash yellows. Not a lot is known about this disease. It involves phytoplasma—a specialized type of bacteria that feed on the inner bark of the tree. Ash blight is extremely common in my area. I can spot it by seeing populations of trees with dying tops. Trees often die very slowly and in groups. Along their trunks, they make clusters of very weak sprouts. They are not riddled with woodpecker damage as a tree with EAB would have. The wood is totally sound. For the last 10 years, I have harvested a very large percentage of my firewood from ash infected with blight.

## White Ash Wood

White ash grows some of the highest-quality wood in the world. Sometimes it is easy for us to overlook amazing things when they are common. Ash is generally taken for granted, but the wood is totally unique and supports several industries.

One time I was talking to my friend Jeff, who is possibly the most skilled woodworker I have ever met. Jeff manages his own woodlot as well as consulting for other landowners. He was lamenting the loss of ash trees, and said to me, "I just don't know what they're going to make rake handles out of after the ash are gone." I thought to myself, *Who cares about rake handles?* I later realized, though, that rake handles really illustrate the qualities of ash that make it so unique and valuable. We take rake handles for granted. Many people would not care whether they were made of fiberglass or wood, but those people probably don't spend as many hours with hand tools as someone like Jeff does. Ash is used for rake handles because it is so light, strong, and stable. Most types of wood will twist and warp as they dry, especially if they are cut into thin dimensions. There are virtually no other woods that could remain stable as a thin handle of a rake.

White ash is not a fancy-looking wood. It does not have the color of cherry or the rays of oak. It is not rot-resistant like locust. And yet it is simply amazing wood, growing as fast and as straight as pine, but still possessing high fuel value, easy workability, and a subtle beauty.

Demonstrating its ability to grow a renewable resource, this ash tree was cut to the stump one year prior to this picture.

Ash is quality. It is very light and very strong at the same time. I have used it for making longbows with draw weights well over 50 pounds. The bending and springing of wood is a tremendous strain. For wood to be able to bend and snap back like that is no easy task.

Ash is most famous for its use in baseball bats, where it truly excels. Maple or hickory would be just as strong, but they are also much heavier. That combination of strength and lightness is owned by ash alone.

Today ash lumber is finding its way into cabinets, trim, furniture, and flooring. Its subtle, natural beauty is only recently being appreciated, oddly enough at a time when the trees are dying by the millions.

White ash is the greatest firewood in the history of firewood. It burns hot, with a BTU value similar to white oak and sugar maple. It is the easiest wood to process. Trees grow straight and with few branches. The rounds split as if by magic with a good whack from a maul. I have split a lot of wood over the years, and there is nothing so satisfying as white ash. Ash not only burns hot and is easy to process, but also seasons faster than other woods. It is considered one of the only woods that you can burn green. (Still, in my experience, wood always burns a lot better when seasoned.)

## Wildlife Value

I used to think that ash trees had virtually no value to wildlife, but with a more observant eye, I see that's not true at all. Ash seedlings endure extreme levels of browse. They are one of the most prolific seedlings in many forests, thickets, and old fields. I have seen them chewed on repeatedly by deer and rabbits year after year. Some trees are 15 years old and no taller than a foot or two. I believe ash serves as a major food source for many deer in my area.

Ash trees grow fast and can become enormous. Just growing all that wood creates habitat for roosting birds, mammals, woodpeckers, and countless tiny critters. Old dead ash trees grown out in the open sun are often covered in vines and bushes and full of holes the size of a raccoon.

Ash do not cast a dense shade. The understory of an ash grove is always a tangle of shrubs, vines, herbs, and other trees.

Ash seeds rain down in cycles. They generally fruit every few years and usually all together. When the ash has a big crop of seeds, they are an important food for numerous songbirds, gamebirds, and mammals large and small. Ash bark makes up 5 to 10 percent of beaver diets in the eastern half of the US.[3]

Dozens of species of Lepidoptera (butterfly and moth) larvae feed on ash leaves. These Lepidoptera are an important food for birds and are also important in their own right. With the passage of ash trees in the US, countless species of Lepidoptera and other insects will go extinct. Many insects are so specialized that their life is tied to a single species. Some of the insects that feed on ash feed on nothing else. When you lose a tree, you realize there was an entire world connected to it. The tree was not just one being, but an ecosystem of creatures and strange organisms like lichens, phytoplasmas, fungi, tiny bugs, and a host of others that we are not able to perceive.

## Commercial Possibilities

Timber and firewood are the traditional ways people have made money from ash trees. Other possibilities include bark baskets, splint baskets, and seed collecting. Biomass production on a small scale, such as a biochar business, seems feasible to me if you have a lot of ash nearby. They coppice so well and they are very enjoyable to work with. It is a lot easier to drag piles of ash branches around than most anything else.

I imagine lumber riddled with the tunnels of emerald ash borers will one day be very valuable, as it will tell the story of this tree.

I hope that I am not saying goodbye to ash trees. It is hard for me to imagine them disappearing. Many times I have been struck by the sight of their winter branches stretching into vibrant evening skies. I hope that if I am an old man and they've disappeared from here, I have saved a big jar of seeds and am telling some kids about the time I almost got killed felling a giant ash. Or maybe I'll tell them about the ash that was undercut by our creek. It was a 100-year-old giant. You could walk under half the root system into a mini cave under the bank. The exposed roots held on to a perfectly smooth stone half the size of a volleyball. They had formed around the stone, creating a round wooden nest. I hope that I tell stories of ash trees and I hope that other people do, too.

CHAPTER ELEVEN

# Mulberry
## The Giving Tree

*I*magine a large spreading tree that has delicious fruit hanging from every leaf node. The fruit are various colors and resemble blackberries. There appear to be millions of them on a single tree. Every day all summer long, new berries ripen. That is the mulberry. I believe it might be the most generous tree on Earth. There are trees like the apple or the oak that will drop a huge crop of food, but a good mulberry tree will have a continuously unfolding enormous crop for months on end.

While some adults hate mulberry trees, virtually every songbird and kid loves them. They are such generous trees that in tropical regions,

Overwhelming abundance.

mulberries are capable of bearing fruit 12 months of the year. Here in the Northeast, we can find individual trees that will ripen their delicious berries from June until the end of September. These trees are capable of holding unfathomable amounts of fruit every year, and that is one reason I call the mulberry a giving tree.

In some circles the mulberry has a reputation for being invasive. This is an unfortunate accusation, particularly here in New York State. White mulberries have been here for centuries, and red mulberries are native. They have had ample time to become invasive. However, I have never seen more than a few scattered trees in one place. In the Northeast they do not form dense thickets like Norway maples or ailanthus trees. A field left unmowed is about a million times more likely to fill with honeysuckle, European buckthorn, autumn olive, and multiflora rose than with mulberry trees.

Mulberries can be anywhere from 6 to 70 feet tall, and they are often just as wide as they are tall. In general, they have similar shapes and growth habits to box elders. The berries vary in taste as much as in color. Some mulberries are just not that good, while others are outstanding. Don't judge the entire genus based on a couple of taste tests. Excellent-flavored trees are out there. I have found several great ones in the wild, including one individual that is better than any cultivar. Mulberries look like blackberries, but are often much sweeter. It is almost impossible to find a fruiting mulberry tree that is not covered in songbirds. They are living bird feeders and true bird magnets; I have never seen a more reliable way to attract orioles than to have a fruiting mulberry around. If mulberries are near waterways, there will always be ducks and geese underneath. They also attract turkeys, as well as squirrels, chipmunks, and just about every other mammal.

These spreading trees often have low drooping branches for easy picking. The dark fruits stain fingers and tongues, and purple bird poop stains cars and sidewalks.

Mulberry trees can grow just about anywhere, and fast. They grow out of the cracks of sidewalks as well as in rich floodplains, vacant lots, and just about any open area. They will germinate in the shade and sometimes fruit there. They can tolerate flooding, drought, and unreasonable pH levels at either end of the spectrum. Mulberries grow fine under the full shade of black walnuts and still produce decent crops.

They are the epitome of a resilient ally, a tree that will provide despite the worst hardships and setbacks.

There are no poisonous look-alikes to the mulberry. Almost every city in the United States has some growing wild.

## Species

Four species of mulberry can be found in North America. They are confusingly named by colors that seem to have little significance.

### White Mulberry (*Morus alba*)

The white mulberry is native to Eurasia. It produces berries that can be white, but are more often red, black, purple, or lavender. The leaves are the primary food source for silkworms in Asia, and the tree has been domesticated for thousands of years in China. This was attempted in America during the early colonial days up until the mid-1800s. The US silk industry never amounted to anything, primarily because of cheaper overseas labor. One tragic legacy of this episode, though, was the accidental introduction of the gypsy moth, which was being bred as a possible silkworm.

The silk industry may never have taken off in the US, but the trees did. *Morus alba* has been very successful at naturalizing itself across the country. It is now found growing wild in all of the Lower 48 states.

*Morus alba* is a very tough tree. It will be quite happy growing out of a sidewalk crack or really almost anywhere besides wetlands. It is extremely drought-tolerant and is cold-hardy to zone 4, with a few populations surviving into zone 3.

*Morus alba* is considered a threat by many native-plant enthusiasts, not because it crowds out the native mulberry, but because it hybridizes with native mulberries readily. *M. alba* does extremely well in urban environments, and that is where you are most likely to find it growing. Some of its most common sites are along highways, in vacant lots, and up against buildings.

The leaves of white mulberries can be all sorts of shapes, even on the same tree. They can have weird irregular lobes, or just be heart-shaped.

There are many forms of *Morus alba,* including weeping, contorted, dwarf, and giants reaching well over 50 feet tall. Countless cultivars of *M. alba* and its hybrids have been selected for fruit and form. Tragically, many cultivars are fruitless male trees. Sometimes I just don't understand people. Why would anyone want to select for a fruitless mulberry tree? That's like having a car that doesn't drive or a pen that doesn't write.

### Red Mulberry (*Morus rubra*)

This tree is native to the eastern half of the US. It has become rare due in large part to white mulberries crossing with it. Misidentification of *Morus rubra* is rampant, and virtually all nurseries and botanical gardens that claim to have *Morus rubra* actually have *M. alba* or a hybrid. For a detailed discussion of this catastrophe, check out *Incredible Wild Edibles* by Sam Thayer. His chapter on mulberries does an excellent job of explaining how *M. rubra* has become virtually a lost species through misidentification.

*Morus rubra* makes red- to black-colored berries and can grow up to 70 feet tall. After learning from Sam Thayer, I believe I have only actually found one *M. rubra* tree in my life. I previously thought I had seen hundreds. The leaves are unlobed on mature specimens and resemble basswood. Again, check out Sam's book if you really want to know about properly identifying this tree. This mistake permeates the horticultural and restoration industries.[1]

### Black Mulberry (*Morus nigra*)

This is the mulberry tree that is most cultivated around the world. Unfortunately for me, it is not very cold-hardy. Zone 7 defines the limits of its northern range. Black mulberries are considered the best-tasting fruits and are the largest. Some cultivars have berries that are 4 inches long! They have been cultivated in Central Asia for centuries, where they are used for juice and dried berries. Black mulberries are considered a staple crop in parts of Afghanistan and Pakistan.

### Texas Mulberry (*Morus microphylla*)

This small tree is also known as the little leaf mulberry. Its native range is in the western two-thirds of Texas. It bears red to black berries.

## Wood

Mulberries can be cut down again and again, and keep sprouting back. They are one of the most resilient species for coppicing. Milky sap oozes from their wounds, just like on a fig tree. The roots are bright orange, and their rot-resistant heartwood is bright yellow when first cut, darkening to a beautiful orange-brown with age. I still remember the first time I saw a mulberry tree cut when I worked for a tree service. Ribbons of yellow sawdust as bright as a highlighter came flying out of the trunk. I brought some of the wood home and made several spoons and cutting boards. As the wood was exposed to light over the coming months, it turned to a deep orange-brown with a beautiful finish.

Mulberry wood is very hard and strong, with uses ranging from fence posts to cutting boards. It is fairly rot-resistant and extremely beautiful. Mulberry is hard and dense and makes great firewood, too.

Mulberry wood is not recognized for its economic value, but its density and rich color make it a gift to specialty woodworkers.

The bark of mulberry stems strips easily. It is a very strong fiber with multiple survival uses. Most tree barks will not strip unless the sap is flowing in spring or early summer, but mulberry offers its fibers any time of year. It is strong enough to use for bow drill fire making without twisting into cordage, or as an emergency dog leash.

## Propagation

Mulberries have been one of my favorite trees for a long time. I have always been so amazed at how much fruit they are capable of producing. I love that the fruit can leave a purple stain on fingers, mouths, and cars; that it can feed a million birds and still leave huge amounts of delicious berries for the rest of us. After a lifetime of generously dropping copious amounts of fruit from the sky, a mulberry tree will leave behind a trunk of beautiful rot-resistant wood. I am intrigued by the commercial possibilities of this tree, from dried fruit to jam to poultry feed. And I am totally amazed that mulberry trees are not widely available at nurseries along with apples and peaches. When I first tried to buy a mulberry tree, I could not find one anywhere locally, so I decided to grow my own. That was in 2007, and I have grown thousands since. Here are some of the lessons I have learned along the way to successfully propagate mulberries. I am not done discovering the nuances of propagating mulberries.

## Cloning Versus Seed

Mulberry trees are dioecious, meaning that some trees are male and some female. If a female tree is growing with no males nearby, the tree will make lots of fruit with either no seeds or unviable seeds, while a male tree will never fruit. Growing trees from seed means that you don't know what sex the seedlings will be until they flower. A seedling mulberry often takes from 5 to 10 years to begin flowering. A grafted mulberry will begin flowering and fruiting its first or second year because it skips the juvenile phase by using the mature wood of an older tree.

If you grow a fruit from seed, it will produce a tree different from its parent, just like every kid is different from their parent. Some mulberry trees' fruit taste much better than the average wild tree. In order to get another tree with the same fruit, it would need to be cloned. It could be grown from cuttings, grafted, or layered. If it was grown from seed, it might be a male or a female and it might make fruit very different from its parent.

The benefits of seedlings are that they are easy to propagate in large numbers and they increase the genetic diversity of the species, allowing mulberries to continue evolving. Clones, while rewarding to grow, are an evolutionary dead end. Seedlings are also the only way to discover new varieties. Seedlings are highly suitable trees for wildlife and livestock, and they often produce decent fruit. The first three seedlings I grew to fruiting age began bearing at age six. Each tree made hundreds of berries that year. One tree had red berries, one black, and one white. They all tasted good. It was a very satisfying experience to pick them with my kids that summer. I believe there is tremendous value on either side of this coin. Growing mulberries from seed and growing them by cloning are both worthwhile pursuits and a great use of time.

## Cuttings

Some varieties of mulberry are much easier to root than others. I have been able to root hardwood cuttings when they are gathered in late winter and placed in trays with bottom heat, keeping the tops cool. I use a very light mix that is about 70 percent sand and 30 percent compost/soil, or 100 percent sand. Hardwood cuttings have rooted at a low percentage for me. Some batches of cuttings treated in this way had around 25 percent success rate, while others were zero.

Softwood cuttings have been more successful for me since I built an intermittent mist system. The cuttings are placed in beds of sand that have a heating cable underneath. A fine mist is sprayed on their tops every 10 minutes for a duration of 10 seconds. I take the cuttings when they are most actively growing during summer. I cut them into pieces about 3 or 4 inches in length. I cut most of the leaf off, just leaving a small portion attached to the cutting. Under these conditions I've had certain varieties root in as little as two weeks, but some others never seem to root. The biggest challenge to softwood cuttings is overwintering the tiny rooted plants. Overwintering rooted hardwood cuttings is easier because the plants are so much larger and stronger.

Many southern growers report that rooting mulberries is as easy as stabbing a branch into the ground. I believe the reason for this is that certain varieties are much easier to root than others. I have not yet found a northern variety that roots with such ease.

## Grafting

Grafting is one method that works for cloning superior female varieties. Mulberries are a little more difficult to graft than apples or pears, but certainly doable. Bench grafting mulberries in May has worked somewhat well for me. I have had about 65 percent take. The real struggle has been overwintering grafted mulberries.

The graft unions on mulberry are tender. They need to be protected for their first year or two in my climate (zone 5). Burying the graft union is sometimes enough, but often not. The most reliable way to overwinter them is to place them in an unheated basement or attached garage and then plant out in the spring after danger of frost has passed.

If you have never grafted before, I think it is a good idea to practice on something else. You can make grafting cuts on any freshly growing shoots of a nearby tree or bush. Save your valuable plant material for when you feel confident. Mulberry bark shreds and peels so easily that you need a very sharp knife and a sure hand.

## Layering

I am currently experimenting with layering mulberries in stool beds. So far it appears they need two growing seasons to root. I'm just getting

Mulberry seedlings are easy to start in great numbers, but watch out for slugs!

started with this method and am excited about its possibilities. Layering offers strong plants on their own roots and can give difficult-to-root varieties plenty of time.

## Seed

Growing mulberries from seed has been very reliable for me. It is easy to start thousands of trees in a few minutes if certain things are given attention. The first thing to be aware of is the tree that you gather seed from. Some fruit-bearing trees do not have a male tree nearby, so any seed you find in the fruit is likely to be sterile. It is best to find a fruiting mulberry that has lots of close neighbors. This will bring seed viability to very high levels.

Seed can be cleaned from the fruit and dried. I obliterate the fruit in a bucket of water with a paint mixer. I then pour off the excess water, taking care not to dump the seeds. Add more water, stir it, dump water, add water, stir, dump, add water, stir, dump . . . Eventually all the fruit pulp will have floated away and you will be left with a considerable amount of seed in the bottom. This method of cleaning seed is known as decanting, and it works well for separating many types of seeds from fleshy fruit.

Mulberry seeds do not require stratification or treatment of any kind. I store them dry as you would vegetable seeds. I lightly rake them into the soil so that they are at or just below the surface. They usually sprout within a couple of weeks.

One catastrophe to avoid when growing seed is slugs. They can easily wipe out an entire tray in a single night. Mulching with fresh sawdust has helped me a lot with this. There are many people out there with lots of strategies for dealing with slugs. Whatever you do, just beware of them. I cannot overemphasize how quickly they can devastate a thousand mulberry sprouts.

# Mulberry

Two-year-old mulberry seedlings dug up from the nursery.

Another hazard for baby mulberries is frost. I wait until danger of frost has passed before sowing seeds. If a frost is coming, be sure to protect them or they will be killed by the freezing temperatures.

Mulberry seedlings often grow very fast. It is almost impossible to stay ahead of them by increasing pot sizes. Trees grown bareroot always have better root systems and are much bigger. One to 3 feet is a typical size for first-year mulberries; second-year trees can be 6 feet with beefy roots and thick stems. Potted trees rarely grow more than 10 inches their first year and are usually horribly root-bound.

Growing your own mulberries from seed or through cloning is very satisfying and rewarding. Trees can begin bearing fruit at a young age and continue to do so for the rest of your life. Mulberries are a gift to yourself, your kids and grandkids, songbirds, wildlife, livestock, and the world in general. There really are very few trees that can match the generosity of a mulberry. So plant a lot; the world is only better for their presence.

# Harvesting and Processing Mulberries

This guy doesn't mind getting messy gathering mulberries.

There are trees from which the fruit needs to be picked and there are individuals whose fruit falls freely. The free-falling trees are much easier to harvest large quantities from. Place a clean sheet or a tarp under the tree, climb, and shake. It's amazing how fast you can fill a bucket using this method. It takes a minute to pick out leaves and twigs from the tarp, but it's still far faster than picking individual berries by hand.

The berries can be made into juice or jam, or dried. If you're making jam, then put them in a food processor to help make the short attached stems unnoticeable.

Dried mulberries are a great addition to cereal or trail mix, or just eaten by themselves. You can dry the berries whole in any dehydrator. I love chewing on them in the winter.

Reportedly, eating unripe mulberries can lead to an interesting experience involving hallucinations. I tried this several times with zero effects. I'm not sure where this information comes from, but I'd love to hear from someone who has personally experienced tripping on unripe mulberries.

# Varieties

There are many varieties of mulberries cultivated in the US and around the world. Here are a handful of them.

**Weeping:** A few different cultivars of mulberry have a weeping form. Many cities plant weeping mulberries that are fruitless male clones. Fortunately, there are also weeping female mulberries that bear small, tasty black berries. Weeping mulberries can be staked to their desired height and then allowed to drape down. I can't think of a more awesome landscape feature for a kid than a tent of fruiting mulberry to crawl into.

**Dwarf Giraldi:** This variety only reaches 6 feet high at maturity. It can be grown in a pot on a patio or balcony. Bears good black fruit.

**Kokuso:** *Morus alba.* An excellent variety from Korea, Kokuso bears delicious large dark purple berries that ripen over a short window (for mulberry), about three weeks in early summer. The berries fall freely from the tree with a light shake, making collection fast and easy. Kokuso is also very hardy, reportedly to zone 3. The leaves on this tree are enormous, about twice as big as a normal mulberry. The growth habit of this tree is somewhat shorter than seedlings. Kokuso has good potential to contribute to a commercial mulberry orchard because the harvest window is condensed, and the berries are of a high quality.

**Illinois Everbearing:** Hybrid of *Morus alba* and *M. rubra.* This hardy tree has a very long ripening window, as berries turn dark and sweet all summer long. The fruits are big and tasty. They need to be picked, as they cling to stems strongly.

**Pakistani:** *Morus nigra.* Hardy to zone 7. Pakistani bears a huge tasty fruit that can be 4 inches long.

**Everloving:** Possibly a hybrid of *Morus rubra* and *M. alba.* I discovered this tree growing wild in Ithaca, New York. It is hardy to at least −25°F (−32°C). The black berries have the best mulberry flavor I've tasted by far. Fruit begins ripening in late June and continues into September. It bears consistent crops from year to year, and the fruits shake down easily. The berries are not as big as Illinois Everbearing, but the flavor and ease of harvest more than make up for that. I have been observing this tree closely for the last nine years and I believe it is worthy of being widely planted in landscapes and orchards. To date, it is my favorite variety.

**White:** There are many varieties and wild trees with white or white/lavender-colored fruit. The ones I have tasted are sweet with little other flavor. I don't enjoy eating white mulberries nearly as much as most dark ones. A good mulberry will not only be sweet but also have a complex flavor. White mulberries are often planted because the fruit will not stain fingers or sidewalks. One thing I do like about white mulberries is that they are very good dried.

## Commercial Possibilities

I believe the mulberry holds serious possibilities for the entrepreneur. There are numerous cultivars available, and wild trees are plentiful for people looking to sell fruit raw or processed. Some of the biggest drawbacks are stems and a short shelf life. I have yet to find a tree that drops the berries without a stem attached. The stem doesn't bother me, but it might be a tough sell to the public. The shelf life of fresh fruit is pretty short, making transportation difficult.

As with most fruits and nuts, I think these deficiencies in mulberry fruit can be overcome through processing and creativity.

Mulberry juice, jam, and dried berries are already commercially available. I cannot even imagine the juice yield per acre in a mulberry orchard. It must be enormous.

U-pick mulberry would surely attract attention. Harvesting mulberries is great fun for kids and families.

One ignored but practical avenue for the mulberry is poultry and livestock feed. Trees planted in or on the edge of pasture will supplement feed for the entire summer when they drop their fruit. The leaves are also excellent feed for cattle, sheep, and goats.

Of all the businesses I have dreamed of over the years, I believe growing mulberries may be one of the best. There is such a mystique around this tree. It is amazing to see people's reactions when I talk about mulberries. Sometimes people are quite turned off to anything mulberry, but more often their eyes light up. I know that a U-pick mulberry orchard would be worthy of a festival drawing folks in from out of state. You just cannot overestimate the power of a tree that holds childhood memories for so many people.

If ever there was a tree to love—a tree so generous and fun that every neighborhood, chicken yard, hedgerow, animal pasture, and old field needs at least one—it is the mulberry.

CHAPTER TWELVE

# Elderberry
## The Caretaker

Rambunctious, energetic, and never offended, elderberries are powerful allies to wildlife, foragers, medicine makers, and gardeners. Elder plants have been used for millennia by people. The uses for elder are as varied and unique as the people who use them. Elderberries are extremely vigorous shrubs growing in thickets, along roadsides, in wetlands, at forest edges, and under openings in the canopy. They are shade-tolerant, but love the sun.

## Species

There are 42 species of elderberry around the world; 12 in North America.

### American Elderberry (*Sambucus canadensis*)

This species is also known as black elderberry. Native to the eastern half of North America, *S. canadensis* is found in a wide variety of habitats. It is a vigorous shrub producing copious amounts of shoots and berries. The tiny, dark berries form in large clusters, ripening toward the end of the summer. *S. canadensis* grows in wetlands, on roadsides, in forest openings, in barely moist to wet fields, and around old homesteads.

### European Elderberry (*Sambucus nigra*)

The European elderberry is considered the same species as *S. canadensis* by botanists, but I have to list it here as a different species. It used to be considered separate. Botanists are constantly changing the categorizations of plants. In reality, the two plants are different enough from each other that *S. canadensis* will thrive here in New York State, while most cultivars of *S. nigra* will languish. *S. nigra* is less vigorous, produces fewer suckers, and is less cold-hardy. It grows less as a bush and more as a small tree.

### Sambucus caerulea

This is from the western US and makes a large tasty blue berry.

### Sambucus racemosa

Native to eastern woodlands, this elder makes a toxic bright red berry. It's not as common as *S. canadensis*. I usually find this plant scattered about in light shade.

# Healthy Productive Stems

One-year-old shoots from well-established plants can sometimes reach 8 feet tall in a single season. Many plants will produce flowers and fruit on new wood that is only a few months old. That is totally amazing to me: a stem rising out of the ground in the spring and winding up taller than me and covered in fruit by the end of the summer. How can a stem be that vigorous and productive? I don't know; it's amazing what elderberries can do.

Elderberries in June responding to being cut to the ground over the winter. By the end of summer, these plants were 9 feet tall with almost as wide a spread and loads of fruit.

# Elderberry

As elderberry stems grow older, they begin to decline. After roughly five to eight years, elderberry stems will start to die. If they are not replaced by new shoots, then the whole bush can go into decline. Wild elderberry bushes rarely live past 20 years of age, but with human help they can live for much longer.

The reason elderberries can live longer with people is because we can cut them down. This may sound counterintuitive, but that is what most shrubs really need. Most shrubs have evolved to grow in open areas with abundant grazing and browsing animals. They are rejuvenated by the disturbance of a large herd of herbivores or by fire. They come from a time when they would regularly be trampled by elephants, mammoths, aurochs, and bison. When old stems die, there is new room for young shoots.

I cut most of my elderberry plants right to the ground every year or every other year. The timing is important. I cut them in the winter when they are dormant and most of their energy is stored in their root systems. In spring the plants flush an abundance of vigorous canes. These canes will reach 8 or more feet tall over the summer, branch, and produce copious amounts of flowers and fruit.

It's important to note that some individuals will not flower on first-year wood. Second-year canes are generally the most productive. Second-year stems are more productive than first-year stems because they will have more lateral branches.

## Wood

Elderberry stems have a soft pith in their centers. Weak, brittle wood surrounds the pith, which is similar in texture to Styrofoam. This pith can be punched out with a thin round file or a large nail. Taking out the pith will leave you with a hollow stem, which has multiple uses. I have seen friends turn them into flutes, my wife has made crayons with the kids, and I have used them as maple spiles.

Apparently the wood of elderberry is poisonous when fresh. So if you are going to use elder stems for flutes or maple spiles, let them dry before using. I've never heard of anyone experiencing any negative effects from using elder wood for these projects, and I've personally drunk a lot of maple sap collected from elder stems.

## Flowers

Elderberry flowers alone make this plant worth growing.

The flowers of elderberry come in great big white clusters starting in early July and sometimes continuing much later into the summer. By blooming late, elderberries reliably miss typical spring frosts that damage plants like apples, peaches, and plums.

The flowers are a magnet for pollinating insects. There is a huge flurry of activity around the blooms.

They are also edible and medicinal. Elderflowers are eaten as fritters, and they are tinctured and/or dried to make a powerful immune-boosting medicine.

## Fruit

Elderberries are tiny, dark berries with a unique flavor. I don't think of them as a berry in the sense that I would want to go out and stuff handfuls of them into my mouth. They have a pretty strong wine-like flavor. Elderberries are best not eaten fresh, but processed into one of many possible products.

They are more medicine than food in my opinion, but can certainly be used for both. Many folks bake elderberries into pies; make jam, wine, or syrup; and more. We often add frozen elderberries to smoothies or oatmeal.

Where the elderberry really shines is in its powerful immune-boosting quality. Elder syrup taken regularly at the onset of a cold has kept me healthy numerous times. Some people may regard this as folklore, but I know that it's real and so does the exploding elderberry industry, which is expected to soon surpass echinacea in herbal supplement sales.

**Elderberry**

## Harvesting

Harvesting elderberries is very fast compared to any other berry I've collected. It takes only a few minutes to gather several pounds. I have heard people bemoan the tediousness of picking elderberries. This is just based on a lack of information. Elderberries come in great big clusters that are easily snapped off the bush. The berries are attached to thin stems in these clusters. To de-stem the berries, place the cluster in the freezer for 20 minutes or so. Tap the frozen cluster on a cookie sheet and all the berries will fall off easily. You can then put them in a bag and keep them in the freezer until you are ready to use them. I learned this trick from Sam Thayer's book *Nature's Garden*.[1]

## Propagation

I used to live on a property where the owners regularly mowed down a huge patch of elderberry. Every time the plants would really get going and begin to flower, they would all be knocked down with a brush hog. I felt so frustrated at this that I decided to dig up and save some of the patch, even though I had nowhere to plant it. In the middle of summer on a hot day, I dug up a clump and tried to stuff it into a pot. It was too big for any pot, so I cut it into sections with an old ax and stuffed each section into a pot. Amazingly, they all lived. I planted them a year later on my own property, where they continue to thrive today, 10 years later.

Propagating elderberries is very easy and rewarding. There are numerous methods, ranging from cuttings, to root divisions, to seed. I have the most experience with cuttings. I have used both hardwood (dormant season) and softwood (summer) cuttings. Both methods have worked very well. I like to gather hardwood cuttings late in the fall and plant them out into nursery beds. Each cutting is only as long as the section between leaf nodes. I prepare the cutting so that it has a node on top. No node is needed on the bottom. I don't use any rooting hormone on elder cuttings and usually have 90 to 100 percent success rate with hardwood cuttings; only slightly less with softwood.

Elderberries are very sensitive to being transplanted after they root if you do not wait until they are dormant. Once dormant, they are easy to transplant. If they're in full leaf, they wilt easily and die. To prevent this, cut back at least 90 percent of the leaf matter. This will prevent wilting, and they will have a much higher success rate.

I have also grown elderberries from root cuttings. One-inch fragments of root planted at or just below the soil line in early spring can lead to enormous 6-foot plants in a single season. However, I have noticed that if I take root cuttings later in the season, they rarely sprout above the soil line at all.

I have not grown elderberries from seed, but have heard from a friend that, with stratification, it is easy.

## Planting

Elderberries are not fussy about a planting site, but they will respond to favorable conditions with extreme generosity. Many people think that elderberries love wet soil and mistakenly plant them in muddy, anaerobic conditions. They can survive in places like that, but they will rarely thrive. Give elderberries a rich, well-drained soil and they will explode with growth and flower right away. They do appreciate moisture, but not saturation. High organic matter and deep mulch is the key to achieving this combination.

A hedge of elderberry. These bushes are planted 6 feet apart and cut to the ground every winter.

# Elderberry

Deer especially love to browse elderberry. It can be so severe that elderberry populations are in decline in many areas with high deer populations. If you can protect young plants until they are well established, they can persist and thrive through a lot of browse.

Some individual elderberries are self-pollinating, but many are not. If you are planting seedlings or varieties that need a pollinator, then a 6-foot spacing works out well.

Elderberries are shade-tolerant, and can do quite well in half a day of sunlight or in dappled shade. However, they will respond to abundant sunlight, the way they do to good soil, with excellent growth.

## Varieties

When I first started growing elderberry cultivars I was amazed at how big the fruit and flower clusters were. I had never seen such large berries or anything like them in the wild before. The wild elderberries are just fine for growing. It's not hard to have a family's yearly supply from a few wild bushes. The cultivars offer even more impressive yields. I like to grow both wild and cultivated elders.

- **York:** Originating from New York. Self-pollinating, very cold-hardy, fruits on first-year canes; huge berries and flower heads.
- **Scotia:** Originated in Nova Scotia. Needs a pollinator; very cold-hardy; does not fruit on first-year canes. Scotia does fruit abundantly on second-year wood. It is also incredibly vigorous, beyond anything I've ever seen. Large berries and blossoms.
- **Wyldewood:** Originated in Oklahoma. Hardiness has been fine here in zone 5

The huge berries and fruit clusters of York elderberry.

I planted this seedling elderberry at my mom's house about five years before the picture was taken. It is a self-pollinating abundant producer that we have named Bubby.

upstate New York. The flower clusters are enormous, the berries of average size. It will fruit on one-year wood, but often needs two-year-old canes to be able to ripen fruit in our short, cool summers.

**Bob Gordon:** Originated in Missouri. This plant is incredible. Huge yields per acre, in some trials three times any other cultivar. Berries and flower clusters are good size. The big difference I see with Bob G. is that the berries are delicious, much better than any other I've tasted. I like to eat these fresh off the bush.

**Marge:** Originated in Missouri. Marge is a cross between *Sambucus canadensis* and *S. nigra*. Flower clusters and berries are very large. Supreme vigor and good hardiness here in zone 5. Selected by Marge Millican.

Many more cultivars exist, and people continue to discover more in the wild and to breed new varieties. A lot of the work going into elderberry breeding happens at the University of Missouri. Individuals are also finding plants all the time, as interest in elderberry is growing very rapidly.

## Wildlife Value

There are at least 60 species of birds that eat elderberries.[2] These berries are loved not only by birds but also by over a dozen species of mammals. The bark, leaves, and buds are highly palatable to herbivores. Native elderberries host an abundance of insect diversity. These shrubs bustle with wildlife activity at a high level—on par with apple trees. They are worth planting for bird activity alone.

Elderberries are loved by songbirds and gamebirds, but are often so productive that it's not hard for people to harvest them at the same time. I've never lost a crop of elderberries to birds the way I do with blueberries or serviceberries some years.

Elderflowers are an excellent source of nectar for beneficial insects.

## Commercial Possibilities

There is a world market for elderberry products, primarily syrup and tincture, but also dried berries and wine. Elderberry flowers are also of commercial value. They are sold dried for tea. Elderflowers are sometimes used in cut flower arrangements.

Elderberries are consumed widely in Europe and increasingly in North America. The University of Missouri is leading the way in commercial elderberry research as numerous elderberry farms continue to spring up throughout the country. The potential exists for anyone with motivation to make an income from growing elderberry. The plants are vigorous and easy to grow, with no significant pests and very high, reliable yields.

The elderberry is a sacred plant to many people. They say that elderberries watch over children. My kids love to be around the robust elderberry bushes on our farm. They play under the shade of the plants, gather berries, and make crafts with the unusual stems. Elderberries are loved by birds, kids, tree huggers, foragers, native-plant enthusiasts, permaculturists, right-wingers, and left-wingers. Every farm, homestead, and park would do well to have an elder patch somewhere to watch over us all.

CHAPTER THIRTEEN

# Hickory
## Pillars of Life

Hickory trees are special. They have tremendous character. From the sprouted nut they will dive down 3 feet before the top is 6 inches tall. Hickory silhouettes in winter will give anyone good reason to pause and look at the sky. There is nothing like the shape of their branches zigzagging back and forth toward the radiant veil of infinity.

I cannot count the fall days I have spent gathering hickory nuts with friends, with family, and by myself. It is one of my favorite activities in all the world. Sometimes there are more nuts than ground showing. Among the diverse genus of the hickories you will find some trees that bear great big thin-shelled nuts, and others with the most delicious nut on Earth locked in a thick shell. There are hickory nuts that are best used for milk, and others that are going to be known as the oilnut. The trees all have tremendously strong wood and live much longer than people.

## Species

There are several species of hickory and significant variation within each species. Many species hybridize with each other in the wild and under cultivation. I am only describing the few species that I am familiar with (here in the northeastern US). There are several others in different parts of the country and a handful in Southeast Asia.

### Shagbark Hickory (*Carya ovata*)
This is the tree with the classic hickory bark that looks like it is peeling off in great vertical strips. Shagbarks are native to the eastern half of the

United States. They are fairly common, growing well alongside oaks and maples. You can find shagbark hickory along the side of the road, on the edges of swamps, on ridgelines and hillsides. It is a very adaptable species.

Shagbark nuts have prominent ridgelines. The buds are large and blunt, shaped like a jester's crown. When they open in the spring, the unfurling leaves are as striking as any flower. I have found shagbark nuts of many shapes and sizes. Flavor can vary from tree to tree; all are great, but some are exceptional.

Shagbark hickory nuts are considered to be the best-

The bark of mature shagbark hickory.

flavored of all the hickories by everyone who has tasted them. The flavor is without parallel. These nuts can be as small as a marble or as big as a golf ball. The shells are very thick. Many attempts at breeding larger shagbark nuts with thinner shells have been made. The primary focus of these attempts involved hybridizing them with thinner-shelled species like pecan and bitternut. I think maybe the focus should be shifted away from breeding the right tree toward processing the enormous wild crop that already exists. Someone with some good mechanical hands should be able to figure out how to crack and separate hickory nuts efficiently. The raw kernels by themselves are the greatest product any nut grower could offer. Millions of pounds fall to the ground uncollected in my county alone. It is very strange that a food so delicious could be so ignored.

## Shellbark Hickory (*Carya lacinosa*)

Very similar in appearance and growth habits to shagbark. Normally has bigger nuts, usually accompanied by a thicker shell. Shellbark is typically found in rich bottomlands.

### Oilnut, aka Bitternut (*Carya cordiformis*)

I believe an entire book will be written about this tree one day. Virtually all literature considers C. *cordiformis* to be an inferior species. The nuts are horribly bitter when eaten raw. I always thought of them as a joke to play on people by offering a taste. However, I have since learned from Sam Thayer that the bitterness is water-soluble. This is a very big deal for people interested in tree crops.

The shells of bitternut are very thin, similar to the shell of an acorn. The nuts are packed full of meat that has an oil content of 75 to 80 percent. The shells are so thin that whole nuts can be run through a commercial oil press. The oil that comes out contains none of the water-soluble bitter tannins. The flavor is the same as shagbark hickory. Everyone I have offered a taste of the oil to looks up at me in surprise and says something like "Wow" or "Oh my God."

Here is a hickory tree that we don't have to shell to enjoy. Here is vegetable oil raining down from the sky in enormous quantities. A 5-gallon bucket of nuts in the shell will yield ¾ gallon of oil. I have filled 5-gallon buckets of bitternuts in as little as 30 minutes. This tree is offering a tremendous gift, if we only can see it. When a million gallons of high-quality vegetable oil fall on the ground and we ignore it and plow up the earth to grow rapeseed (for canola oil), there is a disconnect. The soil suffers, wildlife suffers, and we do, too. The time is here for us to use hickory oil.

Obviously everyone does not need to have their own oil press, but could we not take a lesson from our great-grandparents, who lived at a time when every town had a mill or a press? If there were just one or two oilnut presses set up in states where these trees are abundant, then people could bring nuts in from all over to collection points. High school kids, retired folks, inspired people of all ages can be a part of this. It seems that we would need machinery to compete with established oils like canola and sunflower, but it's possible that this could be much simpler. This past year I collected enough nuts for 7 gallons of oil in only two days of harvesting.

Bitternuts have smooth bark with a lace pattern. Their buds are flattened and a yellow mustard color. The trees generally grow in rich bottomlands, but can also be found on hillsides and ridges. They often have a striking and beautiful root flare, as the bark is even smoother

# Hickory

Bitternut hickory nuts have a "tail" and no ridges along the sides.

*Left,* Bitternut hickory buds are mustard-colored and their bark is smooth, making them easy to differentiate from other hickories. Pignuts can also have smooth bark, but it's usually more ridged, and they do not have yellow buds like these. *Right,* Bitternut hickory nuts are packed with nutmeat and have very thin shells.

Oil pressed from bitternuts tastes as good as shagbark hickory. Maybe it's time to change the name of this tree to oilnut or something more appealing than *bitternut*.

at the base. The husks do not peel off in segments as with most other hickories; they come off more like a rind. The husks have four ridges that only traverse half the nut.

It has been stated that animals will not eat bitternuts and they offer no wildlife value. From my observations this is not true. Squirrels, chipmunks, and mice will definitely eat them.

## Pignut (*Carya glabra*)

Pignuts have a smooth bark, not shaggy, but rougher than bitternut. The bark protrudes a little and has an interlacing pattern. The husks are not segmented like shagbark, but instead peel off like bitternut. The husks lack the ridges found on bitternut and are usually pear-shaped. The nuts are very difficult to shell, but the flavor is very good. A lot of literature suggests that pignuts do not taste good and are only fit for pigs. This has not been my experience; I think they taste great, similar to shagbark. It's possible that people would confuse this species with bitternut, but the husks, buds, and nuts are quite different. My favorite use for pignut is making hickory milk.

### Pecan (*Carya illinoinensis*)

This is the most famous of all the hickories because it is so widely consumed and cultivated. Many wild pecans have a small nut with a thicker shell. These are still easier to shell than any other hickory because they are not locked inside the shell, but fall out freely when cracked. Most commercial pecans come from wild groves that were thinned out and managed for harvesting.

Many superior varieties of pecan exist. These trees have paper-thin shells and large nuts. Pecans have received a lot of their breeding work and cultivation because they are able to bear nuts at a young age. This allows breeders to make selections and crosses within a single career or lifetime.

Pecan hickories are cold-hardy. Many selections can survive zone 4 winters. However, it is not the cold hardiness that has kept pecan production out of the Northeast; it is the short season and lack of heat. Pecans take a long time to ripen their nuts. Here in the North, the nuts are usually killed by a hard freeze before they are able to mature. Work has been done to find pecans that are able to ripen in shorter, cooler summers, primarily by a few scattered nurseries and the Northern Nut Growers Association.

## The Tree in the Woods

Hickories are a big part of the hardwood forests around here. They are full-sized canopy trees, growing alongside giants such as maple, oak, hemlock, and beech. They occur in pure groves and in mixed forests. They are opportunistic and do well in hedgerows and forest edges where sunlight and weeds are abundant. Hickories put down a strong, deep taproot upon sprouting. They will persist through severe abuse, enduring tremendous browse, drought, and competition.

Hickories cast a shade of medium density. They are not so dark as a maple, but not so light as locust. They will allow shrubs and weeds to grow under their closed canopy.

Hickory trees of all species are a serious magnet for squirrels, mice, chipmunks, and other small forest dwellers with sharp teeth. This in turn benefits hawks, owls, foxes, and the like. The hickory genus also hosts an abundance of Lepidoptera species.

Hickories are irregular croppers. They will rarely make two crops in a row. Usually, hickories will fruit heavily every two to three years or so. Some groves in heavy competition can go six years or more between mast years. I am blessed to have 100-year-old hickory trees nearby that drop a load of nuts just about every other year. This boom-and-bust cycle of fruiting can wreak havoc on rodent populations. In a well-balanced nut grove or forest, several species of nut trees take turns having mast years. Hickories can be one player in a collection of oak, chestnut, black walnut, butternut, beech, pine, and hazel.

## Wood

Hickory is one of the toughest woods on Earth. It burns as hot as anthracite coal, having the same BTUs as black locust. It is a beautiful two-toned wood, with light sapwood and a darker heartwood. Hickory is used commercially for hardwood floors, cabinets, and tool handles.

I have carved several handles and longbows from hickory. When split along the grain (rather than sawn through), it is virtually unbreakable. A green hickory branch can be bent into a full circle easily without cracking. There is no comparable tree in the temperate world that grows such a tough and resilient lignin. Locust may be as dense, but it is nowhere near hickory in strength. Osage orange is stronger, but not by much, and it will never grow giant trunks like hickories do.

## Harvesting

Hickory nuts really are the best-tasting nuts in the world. That is not an exaggeration or an overstatement. There's nothing that compares to a kernel of shagbark. Roughly 70 to 80 percent oil by weight, hickories are very rich, tasty drops of butter falling from above. I think all the species taste amazing raw except for bitternut. Pecan is probably fourth on my list of best-tasting hickories. Shagbark, shellbark, and pignut are the top three.

# Hickory

There are plenty of ways to enjoy wild hickory nuts, and plenty of opportunity. In a decent mast year my family is able to gather a pickup truck's worth of shagbark nuts from four large individuals, and we only take a fraction of the crop. The trees are 80 feet tall and covered in nuts from head to toe. Stacks and stacks of nuts rise up to the sky in a quantity so great that it's hard not to gasp or laugh.

I often hear people complain that they can't gather the nuts because of squirrels. I really don't understand. The squirrels are so small and they are scared of you. Whenever I see squirrels gathering nuts, I grab a bucket and walk over. A person can gather so much more than a dozen squirrels if they have a bucket. Rodents have to run back and forth to store each nut away. Squirrels constantly running across the road in the fall is a sure indicator that nuts are ready. Check and see what they are up to; it is a good way to find treasure trees.

Hickories can be gathered with a nut wizard, but it is much faster to gather them on your hands and knees because there will often be so many in one spot. Bring kneepads if you get sore, or use a nut wizard. The window is short, usually only a couple of weeks total. However, bitternuts can often be gathered throughout the winter in a good year.

Once you find a good spot to gather hickory, you will never forget it.

## Processing

Once hickory nuts are gathered, they need to be husked (hulled) and dried right away. You can wait on husking pignuts and bitternuts, but shagbark and shellbark will mold if the husks are not removed promptly. It is no big deal to husk shagbarks and shellbarks; the husk pops off with fingers easily. In fact, some trees will drop nuts right out of the husk for you. If the husk is stuck to the nut, it isn't ripe. They will never ripen off the tree, so if the husk sticks, just forget it and find ones that are ripe.

You can husk bitternuts and pignuts by peeling the husk off, but this is easier if you either let them dry or let them rot. Dried, you have to husk them one by one, which is not that slow. If you let the husks rot, you can do them in batches. The nuts can be left in sacks with the hulls and some leaves until the hulls start to turn black and soft. You can rub your foot back and forth over the sack to loosen up the hulls. I imagine that as the bitternut oil industry develops, someone will come up with a good method for mechanically hulling them.

## Drying

As soon as the nuts are out of the husks, they need to be dried. Simple air-drying on a countertop or on stacks of screen doors will do. I dry them for about a week before storing. They can be stored at room temperature in onion sacks or brown paper bags for years. We've had hickory nuts here that we ate three years after gathering and they tasted fine, without any rancidity. The key is to store them in the shell. Once they are cracked open, the kernels will age and oxidize. If you want to store shelled nuts, then keep them in the fridge or freezer.

## Cracking

Shelling hickory nuts works best with a heavy-duty nutcracker. I highly recommend the Master Nut Cracker, which is made in Missouri. For years I only used a hammer and a stone. If you're cracking the nuts out with a hammer, hold them on edge. You'll get much bigger kernel pieces if the nuts are cracked along the seams rather than just laid down flat.

The Master Nut Cracker works well for butternuts, black walnuts, and hickory nuts.

The kernels are wonderful eaten raw by themselves. They are also an excellent addition to cookies, cereal, and everything else.

It takes time to crack and shell hickory nuts, but it's worth it. Cracking hickories is great for kids and adults. It helps quiet the mind. Depending on nut size, it's not that hard to fill a cup with kernels.

### Hickory Brew/Milk

Cracking hickory nuts takes time. It is doable, but it does take a while. On some productive winter evenings, I will crack

## Hickory

Cracking nuts on a stone with a hammer is an age-old activity that reminds me of people playing cards. I have seen it keep the most rambunctious kids busy for surprisingly long periods of time.

out a pint of nutmeat in an hour or so. This is great for snacking, but difficult to sustain yourself on. To really consume hickory nuts as a part of your diet and fill your body with their tremendous power, boiling them is the way to go.

I learned about this method after reading about traditional Cherokee uses of hickory. Hickory nuts were staples in their cooking. Families had hundreds of bushels of nuts stored. They would crush and boil the nuts for days to extract the oil that would eventually float to the surface. I use a modified version of this. My family, using hammers or mallets, crushes hickory nuts, shells and all, on a big stone. We then boil the nuts and shells together. The more crushed up the nuts, the better. After they have been boiling for a while (10 minutes to two days is how long we boil for, as we just throw the pot on the woodstove), a lot of the nutmeat will float to the surface and the shells will sink. The nutmeat can be skimmed off and eaten and/or added to other dishes. The broth is a great fortifying drink. It can be drunk by itself, or you can add cocoa, maple syrup, or any other spices you'd like. In early spring we like to make hickory brew by boiling the nuts in maple sap rather than water.

Throw in some chopped black birch twigs and you have the best possible drink from the forest.

Hickory brew has been a game changer for us. It's allowed us to really use and even sell hickory nuts. We've brought the drink to festivals and shared with many friends. It is a welcome, hearty drink on chilly days that will line your bones with the strength of a hickory tree.

Taking this a step further, you can pound the nuts into a powder with a mortar and pestle. Boiling them this way yields an even thicker liquid that is a high-quality nut milk.

## Candy

Any nut can be candied, and they're all pretty good, but candied hickory nuts are something else. In a saucepan over low heat, I add equal parts hickory kernels, butter, and maple syrup. Keep stirring until the syrup becomes thick like taffy and then remove the pan from the heat. Allow it to cool, and you now have the best candy this Earth has ever seen.

## Oil

Bitternut and pecan are the only two hickories that can be run through a press with the shells on. In most presses, you have to crush the nuts to fit. You do, however, need a commercial oil press of significant strength to do this; a home press can only work with shelled nuts. The oil is of exceptional quality and easily rivals olive oil. I add hickory oil to most meals I eat lately. I put it on salad, rice, beans, popcorn—just about everything.

# Propagation

Hickories are most commonly propagated by seed. Cuttings are not a practical method. Grafting is difficult, though it's certainly done. One of the most difficult aspects of propagating hickory is in transplanting young trees. Hickories put down a strong taproot when they sprout; damage to it can often kill the seedling.

Hickory nuts require a cold moist stratification to germinate. They can dry a little bit, but really should be kept moist. In nature, nuts are planted by squirrels under leaf litter. Mimicking these conditions is the key to sprouting hickory nuts. I store hickory seed nuts in damp sand: in bags in the fridge, in buckets in the basement, or in buckets buried outside. It is essential to protect the nuts from being eaten by rodents, which is why I overwinter them in a controlled space.

I plant them out in beds in the spring and keep a vigilant eye out for rodent tunnels. The nuts sprout very late, usually in June or July. It is easy to forget about them because they come up so late.

There are a few methods to avoid transplant damage with hickory. The first and best is to directly seed the nut into where you want the tree to grow permanently. This works well, especially if you can keep weeds under control and protect the seed and then the seedling from rodents and birds. Short tree tubes work well for direct-seeding hickories in the field.

Another option to ease transplanting of the taproot is to raise the trees in pots. Personally I dislike raising trees in pots. They need frequent watering, require potting soil, and end up with strangely shaped root systems, but it's better than not growing them. There are specialized root pruning pots available if you choose to go that route.

A third method is to raise them in air-pruned beds. This leads to a fibrous root system and also makes it easier to keep rodents out of the bed. For a detailed explanation of air-pruning beds, see chapter 4. Air-pruning beds work very well for hickory. Trees will often reach about 6 to 12 inches tall with a strong, fibrous, intact root system their first year. Planted just in the ground, they will usually only be a few inches tall with one straight, very deep taproot.

## Working with Hickory Trees

A few years ago I found some excellent shagbark hickory trees cropping heavily along the street. The nuts were mostly falling on someone's lawn, so I knocked on the door to ask permission to gather. The man who answered the door was glad to have me take as many as I could. He explained to me that he had planted the trees 30 years ago by just sticking some nuts in the ground. That story says a lot. With a few minutes of effort, that man had set something very powerful in motion. If we work with hickory, we can establish groves of trees that produce easy-to-gather large nuts.

Planting hickory nuts is fine by itself, but when we select unique and excellent trees, then we can take hickory nuts to a higher level. People

*Left,* Hicans are a cross between pecan and shagbark hickory. *Right,* Large shagbark hickory nuts like these are out there waiting to be found.

have taken small wild grasses and through selecting seed sources have been able to cultivate wheat, corn, and a host of other grains. When breeding hickory, we can look to the famous pecan hickory tree.

Pecans have been bred to have thin shells and big nuts. They can start bearing in three to five years, so breeders are able to work with them. Most other species of hickory, however, can take 10 to 30 years to make nuts. This has been a big impediment to breeders making crosses through generations of seedlings. Still, I think this is a poor excuse for our lack of attention on breeding hickory. We have institutions and universities that persist through generations of people. There are also countless individuals who would love to leave behind something for future generations, like a planting of hickory trees. Surely the resources to work with hickory exist. What a gift we can kick down the road, if we can be part of the journey toward a thin-shelled shagbark hickory nut the size of a golf ball. The genetics do exist.

Some work has been done making selections of hickory trees. There are grafted cultivars of hickory available from a few specialized nurseries. If we plant the seeds from those trees or from other excellent sources, what new trees might we find? I have found an excellent shagbark hickory growing in a cemetery of Revolutionary War vets.

The tree appears to be as old as the cemetery. It has outstanding form and health and bears beautiful football-shaped nuts that crack out well. Another tree a few miles away makes large heart-shaped nuts that are just bursting with meat. The more hickory seeds that people plant, the more we can discover trees with outstanding qualities. It seems to me that most schoolyards and parks could use at least a handful of hickory trees planted by kids who will see the nuts one day.

## Working with Hickory Nuts

I realize the challenge in breeding a wild tree that can take 30 years to flower. Perhaps it is not just the trees that we need to work with but also the nuts. The wild crop is already here, already planted, and is enormous. It is our failing in not knowing how to use it. The gift is here and we do not know how to open the package. If this civilization can fly into outer space, then surely we can overcome a thick nut shell. We do have the means; it is a question of will and inspiration.

How complicated a machine would be needed to crack and separate kernels from shells? Whoever can answer this question will start an industry around shagbark hickory. You will see the planting of orchards, the tending of wild groves, and the busy hands of countless people harvesting.

## Commercial Possibilities

Hickory trees allow people who want to be nut growers to start without having to wait for planted trees to grow. There is already an abundant wild crop; it just needs to be harvested.

Hickory milk is one avenue that I have pursued. We have brewed it on-site at festivals to avoid needing a commercial kitchen. Sold by the individual cup, hickory nuts are worth quite a bit of money. If I were to take the milk further, I would approach local coffee shops. And to go further still, it could be bottled and packaged like any other nut milk.

Making kernels into candy greatly increases their value and makes the slow cracking process worthwhile.

Wild pecans abound in their native range and can be harvested for free in many places.

I believe the oil from bitternut hickories offers a really great way to make a living with these trees. As enthusiasm for tree crops builds, the oilnut may become an industry, feeding people and the trees at the same time. How many farmers would leave bitternut hickories around the edges of their fields if they knew the in-shell nuts were worth up to $20 a gallon?

Hickory seeds of all species are also of significant value to the nursery trade, particularly for restoration, wildlife plantings, and soil and water conservation. A normal price for seed is around $6 per pound. A healthy person can gather hundreds of pounds in a day.

Maybe you will plant a hickory tree; maybe you'll carve some of that wood, or burn it in your stove; maybe you'll eat hickory nuts in pies and ice cream and all by themselves around a fire. Whatever you do, I hope that you enjoy hickory trees. They are one of the best gifts from the forest to you.

CHAPTER FOURTEEN

# Hazelnut
## The Provider

Powerful and packed with density on multiple levels, hazelnuts are amazing. There are several species of hazelnut around the world belonging to the *Corylus* genus. Most species of hazelnut are shrubs, but some can be huge forest trees. Hazelnuts have been eaten by people around the temperate world for thousands of years. Pollen records show that hazelnuts were one of the first plants to appear behind retreating glaciers in a volatile and raw world. They really are rugged, adaptable survivors.

Hazels will grow in heavy clay or sand. They can tolerate drought, flooding, and a wide range of pH. They do very well on infertile soils and even better on fertile ones. They are one of the easiest plants to grow. Hazels tolerate extreme competition from weeds and even large trees. With all that said, you will get a lot more out of a hazel planting if it is well tended. As tolerant as they are of adverse conditions is also how generous they are in cultivation.

Their root systems are dense, fibrous, deep, and competitive. Stems rise out of the ground at a quick rate and are strong. They will not break from snow loads or even from occasionally being driven over.

It is a joy to work with plants that really want to grow and require minimal care. The hazelnut is right at the top of my

Hybrid hazelnuts.

list of easy-to-establish, rewarding trees. Not only are they easy to grow, but the gifts that they offer are numerous and significant. From the wood to the flowers to the shells, the husks, and the nuts, the hazelnut is a powerful ally.

## Species

Eighteen species of hazelnut are found in temperate areas throughout Asia, Europe, and North America. They grow in harsh and moderate conditions as varied as the Himalayas and Oregon's Willamette Valley.

### European Hazel (*Corylus avellana*)

These grow as shrubs throughout Europe and Central Asia. There are mountainsides entirely covered with European hazel along the Black Sea and the Caspian Sea. They are primarily gathered by hand and brought into collection points, where they are sent to large processing facilities. Around 90 percent of consumed hazelnuts come from Turkey, and this is how they are grown and harvested for the most part—in mountain culture.

Most of these nuts are much smaller than what we see in stores around the holidays. They are used for products like Nutella, oil, nut butters, desserts, and candies.

*C. avellana* cultivars have the largest nuts, often with the best flavor. The husks usually only come halfway down, leaving the bottom of the nut exposed.

Unfortunately, the European hazel is not very cold-hardy (somewhere around zone 5 or 7 depending on variety). It also lacks disease resistance to eastern filbert blight.

Fortunately, the European hazel has a lot to contribute to breeding programs. Other than nut size and flavor, it is also often resistant to big bud mite.

### American Hazel (*Corylus americana*)

This extremely adaptable shrub has a range across almost all of northern North America, extending up to the tree line in Canada and Alaska. It is hardy to at least zone 2, possibly 1 in certain genetic pools. Not only are the plants hardy, but the flowers are as well. Male catkins can survive temperatures down to −50°F (−45°C), and I have seen female flowers fully open survive at −4°F (−20°C).

American hazel grows in old fields, hedgerows, forest edges, and sandy barren regions, especially around the upper Great Lakes. It forms large pure stands and thickets that can extend for many acres. It is a very tough shrub. It is just as rugged as any other shrub you're likely to find growing in an abandoned pasture. It can reach about 10 feet high and wide at maturity.

American hazel has a big husk that completely envelops the nut. This husk is a safeguard against hungry critters. It hides the nut until *after* it is ripe. American-type husks offer a real sense of protection. Sometimes they are 5 to 10 times bigger than the nut inside. When they're green, they're full of a potent liquid that deters critters. The husks have some use in cosmetic applications.

## Beaked Hazel (*Corylus cornuta*)

Beaked hazel has a similar range to American hazel but stretches farther west and north. Here in upstate New York, I have only found it growing sparsely in the understory. It is a smaller plant than the American. The husks also look different: pointed and very sticky with irritating hairs. Sam Thayer recommends rotting the husks off.

## Turkish Tree Hazel (*Corylus colurna*)

Native to southeast Europe and western Asia, this is the largest species of hazel in the world. Turkish tree hazels grow up to 80 feet tall. They are sometimes planted as city street trees and in parks. They are clean, sturdy trees for urban areas. The nuts are sparser than other species. Sometimes they are of pretty good size and sometimes abundant. Often, they can have disappointingly low yields. Almost all the cultivation of this tree in the United States has centered on its landscape qualities rather than nut production.

## Hybrid Hazels

Hazel species hybridize freely with one another. There have been several breeding programs that crossed hazel species. Until recently, the goal was for a plant with disease resistance, cold hardiness, and a large nut.

A lot of work was done in the 1920s to cross American and European hazels. Some of the biggest plantings were in Geneva, New

Nuts from Jeff Zarnowski's selection, Nitka. Jeff has planted several thousand bushes to find this one of exceptional quality. The nuts are large, have very thin shells, and taste amazing. The bush is cold-hardy, extremely productive, and disease-resistant.

York, where they completely floundered. Very few plants showed any resistance to EFB (eastern filbert blight). Some successful plants did come from some growers, including Carl Weschke and Fred Ashworth, among others. These hybrids were decent plants. Building upon their work, several more breeding programs have become well established. There are many lines of hybrids available today with very different plants.

The most notable hazel breeding program is at Badgersett Research Corporation. Phil Rutter (co-founder of The American Chestnut Foundation) created this program. He is a geneticist who speaks of a hybrid swarm. Hazelnuts have a huge number of genes in each species. To maximize the number of genes in a plant, Phil makes complex hybrids of three or more species. And then he grows the plants in huge numbers. A field planted with a hybrid swarm of genetics will reveal unexpected results somewhere in it. There will be plants that express traits previously unknown. This is done to find resistance to extreme weather events, diseases, insects, and the unforeseen. He has even found a few plants that make insect-pollinated flowers that are carnivorous. The hybrid swarm is a theory that aims to build resiliency into a planting during this extreme period of globalization (the movement of insects, fungi, and climate change). Badgersett's hazels have not been bred for size of the nut or thinner shell so much as for yield (pounds per bush). They are being bred as an oil crop to cut into the vast annual tillage and cultivation of soybeans.

Many other growers in the Midwest have been breeding hazelnuts in this way. Plantings are managed as hedges of thick bushes producing lots of nuts that are not very large and have a thicker shell. Once processed for oil, butter, or kernels, it doesn't actually matter how big the nuts are or how thick the shell was.

### Hazelnut

Nitka hazelnuts. Notice the thin shells. Photo courtesy of Jeff Zarnowski.

There are also hybrid hazel breeding programs for a larger nut with a thinner shell. These are scattered throughout the Northeast and Midwest. Jeff Zarnowski of Z's Nutty Ridge has planted many thousands of hazels over the last 25 years and has found two exceptional plants. He calls them P-1 (*P* stands for "promising") and Nitka. The nuts are large and thin-shelled, with excellent flavor. The plants are very productive and disease-resistant.

The Arbor Day Foundation and the US Department of Defense have also been researching hybrid hazels as an oil crop.

Here at my farm, I've been growing as many hazels from as many different places as I can, hoping to create my own hybrid swarm. I plant hybrid hazels as close as 6 inches apart, down 300-foot rows.

## Cultivated Hazelnuts Around the World

Hazels are grown primarily in certain regions. Turkey grows around 80 percent of the world's commercial crop.[1] Ancient bushes grown densely on steep mountainsides are hand-harvested in the Black Sea region. One single town, Ordu, harvests a quarter of Turkey's nuts. It is a lush pocket of land near the Georgian border. The mountains are very steep there, and at some point in history were planted entirely with hazelnuts of excellent quality. In 2013 there was a late-spring frost that wiped out almost every nut, and then it happened again the following year. The world price of hazelnuts rose from $6 to $17 a pound.[2] I believe we should take this as a lesson on the value of diversity, casting a wide net, and having many baskets with lots of eggs.

Italy grows around 10 percent of the world's hazelnuts, and they are quickly expanding.

About 3 percent of the world's hazelnuts are grown in the Willamette Valley of Washington, which has ideal, favorable conditions for the European hazel (just like Ordu, Turkey).

All of these major hazelnut-producing regions grow *C. avellana*. However, a new industry is forming in the Midwest and the Northeast of the US. Hybrid bushes are being grown in hedge systems. Breeders are searching the North Woods for superior genetics, and farms are springing up at a very fast rate that appears to be accelerating all the time.

## Growing Hazelnuts

Hazelnuts are a pretty easy crop to grow if you have good varieties suited to your area. Anyone in the US should use plants that are resistant to EFB.

Different growers use different techniques for spacing and pruning.

In the Pacific Northwest, hazels are grown as single-stemmed trees. Throughout the year, sprouts from the base are burned off with herbicide. Orchards are planted on level ground, or the earth is graded for flatness. The nuts are harvested when they drop from the tree out of the husk and are picked up by machine sweepers.

If this doesn't appeal to you, don't worry: Most other growers in the world allow hazels to be the bushes that they are. You can plant them either in a hedgerow or as individuals. Planting the bushes to stand alone, they should be spaced 8 to 15 feet apart. When planted as a hedge, 2 to 3 feet between plants and 15 feet between rows works well.

Though the plants can tolerate wet ground, they do much better when planted in well-drained soil. Since I have a clay hillside, I plant them on berms.

They can tolerate a low pH, but will do best at around 6.5. Animal manures are an excellent way to keep the pH toward neutral in a

Hybrid hazelnuts growing as hedges at Z's Nutty Ridge, McGraw, New York. Photo courtesy of Jeff Zarnowski.

hazel planting, as they have so many other benefits. Many growers are experimenting with grazing sheep, horses, and poultry down the aisles between hazel rows. Sheep have been shown to keep the bottoms of the bushes cleaned out enough that rodent habitat is greatly reduced.

To keep hazel shrubs vigorous and healthy, older stems should be pruned out. You can either remove individual stems that are over 4 or 5 years old, or coppice the entire plant every 10 to 15 years. A hazelnut left unpruned has a life expectancy of around 50 years, but regularly coppiced plants have been known to survive well over 1,000 years.

Hazelnuts will do best in full sun. They are shade-tolerant, but productivity will be greatly reduced.

## Flowering and Pollination

Hazels are wind-pollinated. Each plant produces male and female flowers. The male flowers are dangly catkins. They appear in the fall and persist throughout the winter, finally opening and shedding pollen early in the spring. The female flowers are very small, beautiful pink stars that form on the tip of a bud in very early spring.

Hazelnut male catkins in winter.

Hazel flowers are very cold-hardy. The male catkins are hardy to at least −50°F (−45°C); the female flowers on some individuals are hardy to at least −4°F (−20°C). We have had very volatile springs the last several years in upstate New York. Temperatures have swung into the 80s (27–32°C) in March some years and then back down to below 0°F (−18°C). Many plants have been blooming and emerging early only to get damaged by cold. From my observations, the hybrid hazelnuts suffer little from this. There are differences, though. Trees that are majority European heritage are more likely to lose a crop from frost. Plants with majority American genetics seem to usually come through with a lot of nuts despite the most severe of frosts.

Hazels are not self-pollinating. They also will not always pollinate one another if their genes are too similar. It's best to plant three or more and keep them close together. In Turkey two bushes are sometimes planted in the same hole to ensure good cross-pollination.

## Harvesting and Processing Hazelnuts

There are two questions I hear over and over when I talk to folks about nut growing: "How long do I have to wait for nuts?" and "How do I beat the squirrels?"

Hazelnuts can start flowering as early as their first year on rare occasions. Usually they will start making a few nuts around age three to five. They can be very productive by year seven or eight.

Harvesting hazels is best done while they are in the husk. If you wait until they fall out, critters will get most of them. The bulk of my hazel harvest happens in late August, not in the fall.

Ripe hazelnuts will be fully enclosed in the husk and often still green. To determine if a bush has ripe nuts, I push on the nut. If it can move back and forth in the husk, then it's ripe. If it doesn't move, then

## Hazelnut

it's not ripe. It doesn't matter if the nuts are white/green. If they can release from the husk, then they are ready for harvest.

I collect the whole husks and dry them for a minimum of a couple of weeks. Once the husks are totally dry, they will separate from the hulls much more easily. You can pull the husks apart to get the nuts. If you're hulling a larger quantity, you can put them in a sack and walk all over them a lot, rubbing your feet vigorously back and forth. This will hull a good portion of what's in the sack.

Hazelnut husks vary in size, shape, and color (especially among hybrids). It is satisfying work to gather the clusters of nuts.

There are many growers experimenting with different hulling machines. Most of them look something like a bucket with rubber paddles spinning around that beat up the hulls. These bucket huskers are powered by a drill. The husk pieces can be separated from the nuts by winnowing with a good wind or a fan.

Cracking hazelnuts can be done individually if you just have a few handfuls. If you want to crack loads of hazelnuts, then you will want a good cracker. There are many designs. Some shoot the nuts against a steel plate. Simpler machines involve two rotating ridged plates that pop the nuts open. The Davebilt nutcracker is a simple hand-operated machine that can crack a lot of nuts in a short time.

Once the nuts are cracked, they still need to be separated from the shell. There are two main types of separators. One uses air to blow the shells away, but since they are not lighter than the kernel, aspirators are a bit more complicated than a strong fan. Vibrating separator machines are simple but fascinating devices. Nuts are placed on a slanted tray that is shaken rapidly. Kernels rise to the top of the ramp, and shells slide down. They move in different directions based on their specific gravity. There are a lot of kinks to work out in separating nuts from shells with hazels. No one has yet figured out a truly efficient method for hybrid

Pressing hazelnut oil at home with a small press. Oil is collected in the metal cup. The press cake coming out the end is mostly protein and totally dry. It is easily ground up into an excellent flour. It is hard to beat the flavor of freshly pressed oil.

hazels; however, there are dedicated people working on this. If you are a tinkerer and inspired to be part of the tree crops movement, the time is ripe for you. The Upper Midwest Hazelnut Development Initiative and the New York Tree Crops Alliance are two of several organizations working on the details of processing hybrid hazels.

Once you have hazel kernels separated, the possibilities for use are endless. They are great by themselves. The kernels are excellent in candies, especially mixed with chocolate. They can be ground into hazelnut butter. They can also be crushed and boiled to make hazelnut milk.

Roasted hazelnuts are an incredible treat. Roasting brings out the best flavor in the kernels. You can roast them in a stovetop pan or in an oven. They only take a few minutes at medium heat. Some of the native and hybrid hazels are bitter when eaten raw, but roasting them transforms this flavor into a truly excellent one.

Pressing hazelnuts for oil is one of the best ways to use these nuts. With a small home press, they need to be shelled, but with a large commercial press, they can be run through in the shell.

When you press hazelnuts or anything for oil, you will be left with the press cake as a by-product. Press cake is easily ground into flour. This is very high in protein and tastes great. If the nuts have been shelled prior to pressing, then you can add the press cake flour into all kinds of foods: cookies, granola bars, cereals, breads, muffins, and so on. If you pressed them with the shells on, then you have two options for the press cake. If some of the shells were separated prior to pressing but not all, you can still grind the press cake into flour. It takes a strong grinder, but the flour is still good; it just has extra fiber in it. The press cake is also great for animal feed, especially for poultry, who will appreciate the high protein content as well as the grit. I've no idea what percentage of their diet this could make up, but I'm guessing someone will be inspired enough to learn about raising poultry on press cake as more and more hazel farms get going.

## Using Shells and Wood

These two by-products of growing hazels for nuts are valuable. They are actually some of the most exciting features of growing hazelnuts, in my mind.

Hazel shells are very dense. They give off as many BTUs as anthracite coal. Many modern hazel processing facilities power themselves with the shells burned in biogenerators. The shells are also sometimes used as charcoal for backyard grills. The ash from burned hazel shells is extremely high in many trace minerals, and is a good fertilizer. I believe a person could devote their life's work to finding uses for hazelnut shells.

In Northern Europe hazels have been cultivated for centuries for their wood. Hazel canes are very strong and flexible. Wattle fences and even house walls are woven from the stems. The wood is fairly durable and makes a good alternative to bamboo canes for northern gardeners. Plants are cut to the ground on a cycle of anywhere from 1 to 10 years. Hazel wood also burns very hot and can be used as a fuel, for charcoal, or for biochar.

A by-product of pruning hazel bushes is a lot of wood. There are endless uses for this annual supply of carbon. We can use it to build soil or generate electricity or fuel vehicles. Using wood as a gas is not a new idea. In World War II, Russians converted their vehicles to run off wood gasifiers and burned sticks instead of petroleum.

## Wildlife Value

Dense cover and food throughout the calendar year are what hazels provide. These are excellent plants for wildlife. Their dense thickets provide great cover for rabbits, grouse, turkeys, songbirds, raccoons, deer, and many other animals. The catkins are an important winter food for turkeys, grouse, and deer. The nuts are highly prized by turkeys, crows, jays, rodents, and foragers. Of all the different types of nuts I have grown in my nursery, none have been so marauded by wildlife as the hazel. They are nutrient-dense and the animals seem to know that.

The winter buds and summer leaves provide browse to deer. The stem bark is eaten by rabbits. Many species of native insects feed on hazel leaves. What forest edge would not benefit from the addition of a few hazel bushes?

## Pests and Diseases

There aren't too many pests to worry about with hazelnuts, but the following are important to keep in mind.

### Eastern Filbert Blight

This is a native fungus that grows on hazel stems and eventually kills them. Native hazelnuts live very well with EFB. Some growers believe that EFB is mutually beneficial to American and beaked hazels because it prunes out old wood. European hazels, on the other hand, are totally killed by EFB most of the time. Originally the fungus was just in the eastern US, but now it's been introduced out west and is a serious issue for the orchards out there.

EFB can take decades to show up. Many clones have been released that were thought to be resistant only to show susceptibility down the road. The most resistant plants come from trees with American or beaked hazelnut genetics.

### Big Bud Mite

These are microscopic mites that feed on the inside of buds. They destroy flowers and leaves before they emerge, causing loss of nuts and poor growth. One control is to spray with sulfur, but this costs money,

# Hazelnut

takes time, and is antifungal. I think it makes way more sense to select for resistant plants. The native hazels have poor resistance, while European hazels are highly resistant. By growing hybrids, we can find plants that are resistant to both big bud mite and EFB.

## Hazelnut Weevil

These have a similar life cycle to other curculios (weevils). Adults emerge from the soil and lay eggs into immature nuts in the summer; the eggs turn into larvae that tunnel into the nut, emerge as fat grubs, tunnel underground, and then emerge the next year as adults and lay more eggs. This cycle can be broken, or at least reduced, if most of the nuts are harvested, which prevents the larvae from pupating in the soil. Reportedly, raising the pH to near neutral significantly reduces weevil populations. Beneficial nematodes are a possible option. Curculios are a challenge in many tree crops. There are species of curculio that target apples, peaches, plums, cherries, chestnuts, oaks, and many more.

# Propagation

Propagating hazels is not complicated. It is limited to seed and layering for the most part.

## Seed

Propagation of hazelnuts by seed is not hard if rodents and birds can be kept away. The nuts should not be dried out, but a little drying won't hurt them. They require a cold stratification of a few months. Sometimes hazelnuts will not sprout in the spring, but wait an additional year and sprout the following spring. This double dormancy can be a challenge with many tree seeds. To avoid it, I do not plant my hazelnuts

A sprouted hazelnut. Watch out for birds and rodents!

until they begin to sprout during stratification. Sharp changes in soil temperature can cause this double dormancy.

There are also germination inhibitors in hazelnuts that can be leached out to reduce the likelihood of double dormancy. This is not a necessary step, but it does seem to improve germination rates. Before stratifying, I soak hazelnuts in water for a week, changing the water daily.

Mice, squirrels, chipmunks, blue jays, and crows can decimate a bed of hazel seedlings. They will pull up plants anytime during the first season to get the kernel. You can build rodent-proof beds, eliminate habitat around a bed, or have them right next to the house where you can watch them very closely. Hazel seedlings transplant very well once started.

## Layering

Hazelnuts can also be propagated by layering if you have a superior plant worth cloning. They can take two years to root if a branch is just pinned down. If you use a twist-tie to girdle the stem, they can often root in one year. Growing hazels in layered stool beds is how they have traditionally been propagated for orchards. Today they are often being produced in tissue culture.

# Creating Change

Hazels provide food, gather carbon, build soil, increase wildlife habitat and diversity, can be used for windbreaks, and are grown for their wood. These plants add strength to the world. If you care about food justice or climate change or wildlife, then hazelnuts are your allies. They will do the work willingly and very well. These spreading bushes are providers. While annual agriculture supplies us with abundant food, it also erases the landscape every year and burns up the carbon in our soils. Hazelnuts are a force that is far more effective than any vote or dollar in creating the change we want to see.

CHAPTER FIFTEEN

# Black Locust
## The Restoration Tree

Black locust (*Robinia pseudoacacia*) is a tree that is both hated and loved. On one side of the fence, you will find people who say that black locust is a horrible invasive species. On the other, folks will say black locust is a miracle tree with endless uses and ecological services. In the middle, you will find black locust itself, as a post holding up the fence.

Originally black locust was native to central Appalachia and the Ozark Mountains. It has spread from these locations and is now found naturalized in every continental state, in several Canadian provinces, and in parts of Europe. Locust is an opportunist. The trees will find a home where destruction and disturbance occur. They love abandoned gravel pits, vacant lots, roadsides, and transitioning fields. Locusts grow fast and will spread into any well-drained sunny location with shocking speed. They spread by root runners, occasionally extending 20 feet in a single season. Give black locust some space and it will create a beautiful grove in the harshest of environments, or in the most fertile soil.

## Black and Honey Locust

Black locust is not honey locust. The trees are very different. Honey locust is commonly planted as a male, thornless clone in yards and parks. It's in a different genus from black—the genus *Gleditsia*, of which there are several members. Honey locusts spread by seed and only a little by root suckers—nothing like the way black locust does. The thorns on honey locust are big enough to kill a person; they can easily be 6 inches long, with branches of more thorns upon them. On most wild trees, the thorns of honey locust cover the trunk completely. I believe this may be a defense they developed to keep giant sloths from climbing them to reach their giant sweet pods. Which is yet another difference between *Gleditsia* and *Robinia*: Honey locust pods are around a foot long; black locust pods are just a few inches. Black locust thorns are also less than an inch long. The flowers of the two species are different. The wood is also different: Black locust has dark yellow wood, while honey locust is pink. They are different trees.

## Growth

Black locust is an incredibly fast-growing tree. It is able to form relationships in the soil with certain bacteria that allow it to fix nitrogen right out of the atmosphere. This ability allows locust trees to grow in very poor soils, so long as they are out in the sunshine and the ground is not waterlogged. They can thrive in pure sand or on the most eroded clay hillside.

I have seen black locust trees grow 6 feet in their first year of life (though 2 to 3 feet is more typical for starting from seed). Established locusts that are cut down can put on as much as 10 feet of regrowth the following year. Within seven years black locust can be harvested for small-diameter firewood. Within a 20-year span, they are often big enough for lumber.

This is also a tree that will spread and form large colonies. Rarely do we see a single black locust trunk by itself. They send out runners and sprout up endlessly until they reach shade or a barrier (road, water, lawn, et cetera). Many black locust colonies are several acres large. They spread as vigorously into an open field as quaking aspens. Once they get established in a field they will be very competitive, even under heavy browse conditions.

## Ecological Niche

Black locusts will stampede out into a field or old gravel pit, or anywhere that things have been opened up for them. They are a pioneer species that will not become established in a forest. Locusts are opportunistic, looking for places where the canopy has been cleared.

Black locust casts a very light shade. The leaves are made up of small round leaflets that allow a tremendous amount of light to pass through. The shade created by black locusts is so weak that undergrowth is always rampant underneath them. Most stands of black locust are tangles of honeysuckle and multiflora rose. Where exotic shrubs do not dominate the understory, hardwood tree seedlings find an excellent place to become established. The light shade of locust offers protection, while they improve the soil through their nitrogen fixation and easily compostable leaf litter.

Black locusts are short-lived trees. Because of their shallow root system, they typically start falling over by the time they reach 60-plus years of age. By this time an abundance of hardwood seedlings have become established in the understory. The most common tree to become established in a locust grove in the Northeast is sugar maple. Locust-maple woods paint a very clear picture of an abandoned pasture that evolved into a grove of locust and then the maple. Sugar maples have a much harder time establishing themselves in an open field.

Seed production of black locust begins early and can be heavy. The trees produce pea-shaped pods containing a row of small, hard seeds. These edible seeds, with their hard seed coat, can remain dormant in the soil for decades. Perhaps they are waiting for the next forest disturbance to sprout again. The seedpods flutter off in the wind, but they do not travel very far at all. Many trees and plants that rely on disturbed environments will make long-lived seeds that lie dormant until conditions are right. For example, lotus seeds can remain dormant for thousands of years—some have been sprouted from Egyptian tombs. I don't know how long black locust seeds can last, but I bet it is at least several decades.

## Flowering

Strings of white and lavender pea-like flowers hang over the trees as if they are covered in robes of blossoms in late spring. Stands of black locust can be seen from very far away during this time of year. Driving down roads through urban and suburban areas, we can suddenly see just how abundant the locusts are: Bright white trees lie around every curve. The flowers are edible and delicious. I think they taste just like sugar snap peas. They are a special treat for only about a week every year.

## Trees of Power

Black locust flowers.

They have to be picked fresh for best flavor.

Honeybees as well as many other pollinating insects will flock to these fragrant blossoms in great numbers. Black locust is considered one of the biggest nectar flows of the season by beekeepers.

Several varieties of locust are cultivated for their blossoms and planted as ornamental trees. For the most part these are purple-flowered varieties or trees with interesting forms.

## Wood

This is where black locust really stands out and makes a name for itself. Despite its rapid growth rate, locust produces a super-dense wood. It is a world-class hardwood as valuable as most tropical hardwoods and is traded around the Earth as a premium product. Around my neighborhood typical softwood lumber is 50 cents a board foot, while hardwoods like maple and oak can range from $1 to $3, usually. The black locust mill down the road sells lumber for as much as $7 a board foot and cannot keep any in stock because demand is so high.

Locust is one of the most rot-resistant woods on the planet. It is used for boardwalks, bridges, piers, fence posts, furniture, musical instruments, decks, docks, playgrounds, raised beds, and terracing. It is an immensely useful wood, and also incredibly beautiful. Locust wood is easily identified by anyone who is familiar with it. The rings are porous with dark and light alternating bands of deep brown-gold. When it's first cut, locust is bright yellow, and then it darkens with age and exposure to light. A touch of oil brings out the best in locust.

The rot resistance of black locust is better than that of oak, cedar, and pressure-treated lumber. Pressure-treated lumber is created by infusing wood with decay-resisting chemicals at high pressure. This lumber is

## Some Notes on Black Locust Rot Resistance

- The inner heartwood of black locust is rot-resistant, not the outer ring of white sapwood.
- Heart rots destroy the rot resistance of locust.
- Kiln-dried locust has almost no decay resistance compared with air-dried.
- In the ground, posts often last over 50 years.
- Traditionally, the Haudenosaunee (Iroquois) used black locust poles in building their longhouses. They charred the ends that were in contact with the soil to extend the life of the post.

widely used for picnic tables, decks, raised beds, and playgrounds. Unfortunately, it is created by using toxic chemicals that leach out of the wood over time. Black locust lasts much longer than pressure-treated lumber and is safe enough for kids to eat off.

The wood of locust is also excellent fuel. It burns hotter and longer than oak, maple, and just about every other hardwood. It is tied with hickory at 26.8 million BTUs per cord. Because these trees grow so rapidly and can be cut over and over, their value as a renewable fuel is significant. The word *renewable* literally explains consuming locust as a source of heat and energy. The trees are building ultra-dense wood from elements in the sky. Actual soil is not even required, as these nitrogen-fixing trees make their own.

## Pests and Diseases

There are some pests of black locust to be aware of, most notably the locust borer, which is the biggest challenge to producing quality timber.

### Locust Borer (*Megacyllene robiniae*)

This native insect can cause considerable damage to locusts. Adults deposit eggs under the bark. The eggs hatch into larvae that tunnel through the wood. These tunnels can weaken stems on younger trees, causing them to snap off in a strong wind. The borer tunnels also lessen the value of lumber. Goldenrod is an alternative host for the

Locust borer tunneling was severe enough to cause this tree to fall over. The roots will send up vigorous sprouts in response.

locust borer. Open-grown trees suffer more damage than trees with shaded trunks.

### Heart Rot (*Phellinus robiniae*)

This fungus primarily finds its way into locust trunks by way of tunnels left open by the locust borer.

### Locust Leaf Miner (*Odontota dorsalis*)

This insect eats leaves from the inside out, leaving behind a brown skeletonized leaf. Sometimes the damage is so severe that it can be seen from far distances across valleys. The trees usually recover fine on their own and will leaf back out after an infestation.

## Wildlife and Livestock

Black locust supports wildlife populations in several ways. Numerous species of insects feed on its leaves and nectar. The foliage is often heavily browsed by deer, and rabbits will chew on the young bark in winter. Stands of black locust offer excellent cover because of the thick understory that is usually present. I might be in the minority, but I think black locusts add biodiversity. They transform landscapes from piles of bare sand to living biomes teeming with life. Many forests owe their rich soils and species composition to the work of these remarkable trees.

Black locust also has a lot of potential as a supplemental livestock feed. Branches cut in the summer have been fed to cows, sheep, and goats. The leaves are high in nitrogen. For a larger discussion on this use of black locust, check out Steve Gabriel's book *Silvopasture*.

## Propagation

Black locust is very satisfying to propagate. Most efforts are rewarded with huge trees after one season. Seed and root cuttings are the primary methods.

### Seed

The seeds have a very tough coat. This allows them to remain dormant in the soil for several years until conditions are ripe for growth (usually after a massive disturbance). This coat must be weakened for the seeds to sprout. You can either abrade each seed with a file or use a hot-water treatment. To use the hot-water method, bring a pot of water to a boil and then take it off the heat. Let the water cool a few degrees. Drop the seeds into the hot water and let them soak for 12 to 24 hours. They will swell up and look quite different. Plant immediately after this treatment. They usually sprout within a couple of weeks, quicker if the soil is warm.

Seedlings are not frost-hardy. One night of light frost

Black locust seedling nursery bed. These trees were seeds just a few months prior to this picture.

can kill newly sprouted locusts. I wait to plant mine until danger of frost has passed and cover them if the need arises.

Seedlings transplant very easily, even if roots are broken when digging. I have successfully transplanted locusts by ripping them out of the ground with a weed-wrench or chopping them out with a mattock. In some cases 10 percent or less of their roots were left intact. They still survived fine, and often put on several feet of growth.

If seedlings are dug out of a spot and some roots are left behind, the leftover roots will often turn into new trees the next season. Several times, I have dug up a locust only to find a ring of new trees around where the roots were severed.

### Root Cuttings

This is the easiest method for cloning black locust other than transplanting suckers. To find root cuttings on larger trees, start at the base of the trunk. Find a root flare and follow it as far as you can, digging around the root as you go. The best material is as thick as a finger, but you can use fatter or skinnier roots, too. Two- or 3-inch sections work well. Plant them near the soil surface in spring. They can sometimes take a long time to sprout—often until midsummer. Growth can be phenomenal, reaching 6 to 9 feet the first year.

Arrow-straight black locusts towering above the surrounding forest in Danby, New York.

## Varieties

There are many ornamental varieties of black locust selected for their attractive flowers or interesting form. However, there are also many selections of locust that have been made for timber. Most of the timber breeding work has taken place in Hungary, where huge plantations of black locust are grown. They have done such a good job at improving locust's timber qualities that the trees look completely different. Instead of gnarly, twisted trees, they look like telephone poles.

The USDA has also made some selections of black locust for timber, but these do not compare to the Hungarian trees to date.

Many wild populations of locust have outstanding timber quality. You can find these if you keep an eye out for them. Carl Albers is a friend with a passion for finding trees with excellent qualities. Along with selecting many great nut trees, Carl has found some of the best stands of locust that I have ever seen. Anyone who pays attention can find stands the way he does. As Carl says, "Once your eyes are keyed in, you will find them." These wild stands of superior trees offer us a path forward toward cultivating this remarkable tree.

## Is Black Locust Native or Invasive?

If the definition of *invasive* is "a plant that is able to naturalize outside its native range," then yes, black locust is certainly invasive. However, if the definition includes terms like *degradation* or *crowding out of native species*, we see a different story. We can look at the ecological effects of black locust rather than just maps of where it grows.

For the most part, invasive plants have negative effects on the environment for two reasons. They crowd out native species, thereby limiting diversity, and they are not fed on by native insects, creating biological dead zones. Black locust fits into neither of these roles.

It does not crowd out other native trees the way Norway maple or tree of heaven does. In fact, black locust actually improves conditions for native hardwoods to grow, in a similar fashion to another pioneer species, quaking aspen.

Black locust leaves are fed upon by a lot of insects; the flowers are used by many native pollinators; even the wood is bored into by the native black locust borer.

Here in upstate New York, black locust hasn't traveled far to find a home. Other species that are regularly considered native here from the same home range as locust include eastern redbud, Carolina silver bell, pawpaw, persimmon, vernal witch hazel, fringe tree, and sourwood. It does not seem right to list these trees as native and black locust as not if they share the same place of origin. However, that is exactly what has happened. New York State has joined a host of other states that list black locust as an invasive species and limit its propagation and planting. There are groups of well-meaning people who will cut down locusts and paint the stumps with herbicide. This tree has been demonized by native-plant enthusiasts. It grows where native plants struggle, and then improves conditions for them to appear.

Rather than simply looking at the native range of a plant to determine if it is beneficial or harmful, I prefer to look at the actual impact of a species. From what I have seen, black locust is a wonderful tree able to grow in the most abused landscapes. It is useful to wildlife, native-plant communities, and people.

If we are to pursue ecological solutions to the problems we face, then black locust can be a key player. It is an outstanding soil improver, a biomass producer, and a source of nectar and high-quality, renewable timber. Black locust belongs in the hedgerow of every farm that uses fence posts, beams, or firewood. It can be cut again and again. Black locust will never complain about abuse, either to itself or to the land base; it will always respond with rapid growth and curtains of white blossoms. We can label black locust as an invasive plant, or we can recognize it as an ally on our path to a healthier world.

## Commercial Possibilities

The commercial potential of black locust is much higher than most people would expect. Letting a field turn into a grove of locust can provide a huge portion of a family's income. The grove can be put on a rotation, where a certain percentage is clear-cut every year. The clear-cuts will grow back fast and strong, yielding high-quality wood. There is a big demand for locust posts and lumber for several reasons. Many lakes and rivers do not allow for pressure-treated lumber to be used in docks because of chemical contamination. This leaves the options of tropical hardwoods or black locust.

Another reason the demand is high is because organic certification does not allow for fences or trellises to be installed with pressure-treated lumber.

## Black Locust

High-density apple orchard trellised with locust posts at Eve's Cidery. As demand for organic food grows, so does the demand for locust posts. Photo courtesy of Carl Albers.

    As understanding of the dangers of pressure-treated wood spreads, so does the demand for black locust. Gardeners want it for their raised beds and tomato stakes; parents want it for picnic tables; schools and parks want it for playgrounds; conservationists want it for boardwalks across wetlands. At this point, there is such a demand for black locust that a supplier can be all but guaranteed to sell every piece of wood. I have even seen even the branches sold as natural pea trellises. In addition to all this, people will pay extra money for black locust firewood.

    Nursery stock is another good option for harvesting income. Most state nurseries have stopped selling locust seedlings because of their invasive reputation. At the same time, demand for this tree has skyrocketed. Awareness of this tree's amazing abilities and offerings are spreading as fast as its roots do. There are few nurseries that do not sell out of black locust seedlings well before spring.

    Governments and native-plant purists fight this species. It is banned by law. Trees are cut and the stumps painted with poison to kill the roots. I believe these people are doing what they think is right to support native-plant communities. But if they looked closely at this species in the wild—not in any literature, but just seeing this tree as it is in the world—then a different story would unfold. Black locust supports native-plant communities by healing the scars of industrialization. It creates an environment for other plants to flourish and then it leaves. If left on its own, black locust will shade out its own suckers and seedlings. It will live for six to eight decades and leave behind a black soil filled with the roots of native trees.

Black locust may appear a rampant, selfish species at first glance, but it is really a servant of older ecosystems. This tree freely offers abundant honey, wood, soil, and land reclamation. It is a wonderful gift to an abused world.

## CHAPTER SIXTEEN

# Beech
## The Root Runner

*I*f ever there was a tree for us to know, recognize, and use, it is the beech. Belonging to the Fagaceae family, beeches are closely related to oak and chestnut. They are some of the toughest trees on the planet: *Enduring* might be the best word to describe them, but there really is no word. How could you define yourself with a word? It is that way with beech for me. There is so much to say; how will I ever do this tree justice?

I have walked past millions of beech stems in my life. Their smooth gray bark fits to my hand like no other tree. Drought-tolerant, shade-tolerant, and relentless are the growth habits of beech. They offer far more gifts than are commonly recognized today.

Apart from their sheer beauty, which is significant, beeches provide us with a never-ending supply of dense hardwood, wildlife food and cover, excellent nuts, and resiliency. They are an integral part of the forest, and any farm in a suitable climate can benefit from a beech thicket. But not all people love beeches; in fact, they are hated by an entire industry and regularly cut and treated with herbicide as a rule.

Beech trees can live for centuries. They are yet another gift of creation, wrapped in a package of uncountable intricacies that could take a lifetime to unravel. A person could be kept busy their entire life tending to a beech grove while gathering an income (sustenance).

The gifts of beech are not as obvious as they are with some other species like apples or chestnuts, but they are there nonetheless.

# Species

There are about a dozen species of beech found around the temperate world.

### American Beech (*Fagus grandifolia*)
This is the tree I'm really talking about in this chapter. I know it well, for it is one of the most common trees in the Northeast, where it often dominates the oldest forests along with hemlock. American beech is a giant that can reach 100 feet tall with a trunk so wide, a group of kids can hide behind one. The bark does not fissure or flake as trees mature; instead it stays smooth. Beech trunks look like giant gray elephant legs. The bark is so attractive that people often carve their initials into the trunk (kind of like how kids love to smash ice—I guess it's just fun to ruin things sometimes).

They grow on ridgelines and hillsides, usually in the deep of the woods. Beeches are also planted as shade trees in yards and parks.

### European Beech (*Fagus sylvatica*)
Similar to American beech but a little smaller in stature. Can be a very long-lived tree. Native to Southern and Central Europe. Introduced widely for ornamental/shade purposes.

### Oriental Beech (*Fagus orientalis*)
Native to Central Asia throughout the Black and Caspian Sea regions, where it grows abundantly on mountainsides mixed with chestnut. Has resistance to beech scale.

### Ornamental Variations
Beeches have been widely planted for ornamental purposes for centuries, going back to the earliest gardens of Europe. Clones have been selected for making red/purple leaves. These are known as copper beeches. There are also weeping forms of beech. This may be the most beautiful of all the weeping trees. Weeping beeches look like giant monsters to me. They are filled with a sense of character. I'm not much of an ornamental gardener, but I do appreciate the beautiful form of weeping beeches.

## Growth Habit and Ecology

The way beech trees grow in the forest is unlike any other tree. Though they need sunlight to make food, they can get by with very little of it. Underneath the darkest canopy in the driest, most competitive, nutrient-deprived forest, you can find beech trees. They will grow very slowly in this situation, but they will grow and spread. If the canopy opens up, they will race to fill the space. If the canopy does not open up, they will continue to grow and spread anyway and eventually occupy the canopy and the understory.

Beeches spread their roots far and wide, sending up stems along the way. A single beech root may have dozens of tree trunks sprouting from it.

Because beeches are so shade-tolerant, they will come to dominate the most undisturbed forests. The darker a forest canopy gets, the tougher it is for seedlings. Only certain species can survive in this shade, which may persist for decades or centuries, but sooner or later a tree will fall and the canopy will open. Whichever seedlings were able to survive

Cutting down beech trees only makes them stronger. Each tree is like a network of trunks. All the stems in the photo share a common root system.

until that point will be there ready to take up all that open sky. Beech, sugar maple, and hemlock are the best at this, and that is why we see them in the oldest forests of the Northeast. Shade tolerance is one key to beech's survival in the wild. It is not the fastest grower or the most adaptable; it is simply able to put up with anything. The root systems on beech are tenacious. They cover the ground like a net, sending up countless stems. Beech thickets block just about all the sunlight from reaching the forest floor. If trees are cut, girdled by disease, or burned, they will respond by sending up sprouts at an even higher density.

Beeches are well known for their longevity, as they can live for centuries. Beings with such long life spans have a different perspective on time than we do. Their juvenile phase is much longer than ours. Beech trees generally don't start making seeds until around age 40.

## Wood

To the lumber industry, beech sucks. Thickets of small-diameter stems don't make many board feet, so beech is treated as an invasive weed. Stems are painted with herbicide, which is drawn into the root system, killing the whole tree.

Beech used to be more valuable as a lumber tree, but beech bark disease has turned this species into something less profitable to the hardworking people who run sawmills.

The wood of beech is very strong, hard, and heavy. It is fairly coarse, has rays like oak, and has zero rot resistance. It probably rots faster than any other wood, except maybe birch. It is a wonderful hardwood, though, that's used for flooring, cabinets, and endless other projects.

The lumber of beech is of significant value, but it is the smaller-diameter wood that draws me to this tree. It is hard to overestimate just how many beech stems exist in an acre of beech thicket. I perceive limitless abundance when walking through beech thickets.

The wood of the small stems is just as hard as the wood of the giants. It burns very hot with long-lasting coals. Beech stems make excellent shiitake bolts and beautiful crafts. Leaving the bark on only accentuates their qualities.

The term *renewable fuel* is practically defined by beech. Cut again and again and again, they will not cease to send up new shoots of dense hardwood. Deer browse is minimal on beech, so trees can be coppiced without detriment in most locations. (I have seen deer browse on beech aggressively in some places where their population is very high.)

A stand of beech can yield more carbon than anyone will have the time to process. The sheer biomass of beech stands are tremendous because they are so shade-tolerant that they can retain their lower branches. They can do all this work on very little rainfall, surviving upon the driest ridgelines.

## Nuts

If beechnuts were easier to collect, we would see beech orchards everywhere. The nuts are about the size of a pine nut with just as much flavor. They are extremely high in oil, around 80 percent. I used to go on survival campouts where I would make a shelter and gather my food. On one trip, I was particularly hungry and weak. It was late September, cold, and had been raining for days. My rain jacket was torn to shreds by the third day. Keeping thoughts of misery at bay was a moment-to-moment challenge. Exhausted, I walked down a forest road and found some beech branches hanging down low covered in nuts. I gathered as many as I could for about half an hour and returned to my shelter. I sat there in front of the fire, shelling nut after nut until I had a nice pile of about 2 cups of nutmeat. I ate it all at once and instantly felt a surge of energy through my body. That was the turning point in that trip for me, and I went on to discover many other treasures including a 50-acre patch of loaded blackberries. I will never forget that trip or the beechnuts that changed it for me.

Beechnuts are born in small spiky husks. Inside the husk are two triangle-shaped nuts in the shell. The shells are thinner than an acorn and can be peeled away with a thumbnail.

Unfortunately, beechnuts will often be empty and the shells will reveal nothing but air. I believe this is a pollination issue. Beech trees often form dense stands that are made up of a single individual that cannot pollinate itself. At least that's my theory. Often trees will be full of almost all blanks or almost all filled kernels.

The mast cycles of beech can also be very disappointing. Some trees will take as long as seven years between crops, but every two or three years is more common. However, I have found trees growing in yards and parks that will fruit almost every single year. Perhaps it is simply

Beechnuts are one of the most delicious nuts on Earth.

an issue of conserving resources that keeps beech from fruiting more heavily. The biggest crop load I have seen on beech trees comes from two individuals in my mom's front yard that are just below the leach field with full exposure to the sun.

Another factor that may cause beech to make infrequent seed crops is its ability to propagate itself clonally through root suckers. Whatever the reasons, if a handful of people devoted themselves to the beechnut, we could find trees that crop regularly and heavily. We might even grow them very differently than we see them in the wild. They could be grown on a coppice cycle in dense hedges, rather than as forest giants with impossible-to-reach nuts.

If you've ever tasted beechnuts or handled the wood, then you know that this tree is worthy of our attention.

Many people will be turned off to the idea of working with beech trees simply because they take a long time to make nuts. It is fortunate that there are lots of other people who enjoy planting trees for someone who is not yet born. I love the idea of people finding my trees long after I'm dead. I admire the ancient people of Chile who planted monkey puzzle trees for their nuts. These trees also take 40 years to bear fruit, but they may live for 1,000 years after that. With all that said, grafted beech trees can begin bearing fruit at a young age.

## Wildlife Value

Beech grows far into the cold north country. It is often the only mast tree in areas of the Northeast where oak and hickory are absent. More than 40 species of birds and mammals depend on beech.[1]

The wildlife value of this tree can be very high. The leaves support a large number of Lepidoptera. Their constant supply of rotting wood feeds the life in the soil. Thickets of beech provide good cover to animals and birds in open forests. And of course, beechnuts are relished by just about all birds and mammals. The amount of activity on a single loaded beech tree can be staggering. Flocks of birds will be there all day for weeks. Turkeys, deer, raccoons, and everyone else will show up.

Evidence of bears in an area can be seen by looking at beech trees. They will leave scars where they climbed up the smooth trunks and tore branches to get at the nuts. In fact, black bears are so dependent on beechnuts that their reproductive cycles are closely tied to the masting of the trees. In the Adirondacks of New York, black bears breed every other year. Studies have shown that their cycle is timed with that of the beeches. There is even a direct correlation between human-bear incidents and the beech crop. In years when there are no beechnuts, people are much more likely to have run-ins with hungry bears, according to a 2013 study by Courtney LaMere of SUNY ESF.[2]

A few hundred years ago, giant beech trees fed billions of passenger pigeons. The birds traveled from beech stand to beech stand devouring the crop. The flocks were so numerous that limbs would break off under their weight. When they left, the ground would be covered with two inches of guano. Perhaps this shot of fertilizer helped the trees crop on a more regular basis. We will never know, as people who were here before us thought it was a good idea to kill four billion passenger pigeons and leave none alive.

There have been many criticisms of beech trees since the bark disease was introduced. It is said that beech thickets reduce diversity by excluding all other plants. I agree that in beech thickets, just about nothing else grows. However, beech thickets do not extend for mile after mile; they are patches and they can be contained with mowing, grazing, or a change in terrain. They are pretty much limited to ridges and well-drained hillsides. In themselves, beech thickets are home to a plethora of wild creatures including birds, mammals, insects, and fungi. They are a world unto themselves. Any large property being managed for wildlife would do well to have a beech thicket somewhere.

## Beech Bark Disease

This disease is based upon an insect and at least two fungi. It was accidentally introduced into Nova Scotia in 1890 and has been spreading

Typical beech bark disease in the late stages.

at a rate of about 6 miles a year ever since.[3] Beech bark disease found its way into New York State sometime in the 1960s.

The insect involved is a scale insect called *Cryptococcus fagisuga*. It is a tiny bug covered in a protective white wax. Scale insects attach themselves to the bark of a tree and suck out sap. They have life phases in which they move, but for the most part they just stay in one spot sucking sap through a long thin probe. Following beech scale, nectria fungi move in. These fungi find their way under the shield of bark through the many holes provided by the scale insects. Stems are weakened by the scale and girdled and killed by the fungi. The bark on diseased trees will be very mottled and broken up, nothing like the smooth attractive bark that beeches are known for.

The stem that is attacked is killed, but the tenacious roots are not. Most beech forests in the United States have been transformed by this disease. Instead of giant forest trees with massive trunks and an open understory, we have dense beech thickets. As older stems die, root sprouts take over. The disease is pruning these trees so hard, they come back bushier and bushier. Beech thickets offer little in the way of lumber, but they are an ecosystem unto themselves. A gift or a nuisance depending on the eye of the beholder.

It is wonderful to know that resistance to beech bark disease does exist. Around 1 to 5 percent of American beech trees show resistance.[4] This may not sound like a lot, but it is a huge number. Walk through a forest with thousands and thousands of trees, and it is not hard to find a few that are fully resistant. You will know you have found a resistant beech tree if the disease and scale insects are infesting all the trees in the area. In the middle of all these diseased stems, occasionally you'll see a tree with a perfect smooth trunk. These resistant trees are often surrounded by thickets of less resistant root sprouts. Who knows how this disease will play out in the long run? It is certainly not the end of beech trees.

## Propagation

Growing beech from seed is the same as for acorns and chestnuts. Don't let the nuts dry out or animals eat them. They sprout in the spring following a cold stratification period.

Though I have not tried it, it seems reasonable that beech can be propagated clonally by root cuttings. It is not hard to find the roots growing along the surface of the ground. There are many sections where a small vertical stem will be rising out of the horizontal root. These would be the best sections to take.

Growing beech from stem cuttings is extremely difficult.

Grafting is used to clone many ornamental trees and trees in breeding programs. Grafted beeches can start bearing nuts within one or two years because scions are coming from an already mature tree.

## Commercial Possibilities

Nuts seem like an obvious choice, but they are difficult to gather in large quantities in the wild. There are no commercial orchards or institutions cultivating beech for its nuts, but perhaps someone will change that. It would be a shame not to.

I think the real value in beech comes from its endless suckering habit. Stems can be gathered by the thousands in a small beech woodlot. They can be cut repeatedly, and the more you take, the more will grow back. These stems have endless applications. Certainly they could be put to good use by any creative woodworker. However, shiitake bolts offer the easiest avenue for converting beech stems into food and money. There is a growing interest and demand for log-grown shiitakes. Mushrooms raised on sawdust indoors do not have the same flavor and probably nutrients as log-grown ones. Shiitake growers are in the business of buying bolts. These are 3-foot-long sections of wood that they use for inoculation. A stand of beech can supply an endless and renewable source of shiitake bolts.

I used to go on survival campouts where I would make a shelter out of sticks and leaves. These shelters had to keep me warm and dry even if it was 34°F (1°C) and raining hard for days. The amount of leaves needed to make one of these huts is serious. I always preferred to find refuge around beech trees because there is an abundance of sticks and leaves under them at all times of year. These trees sheltered me and filled me with a sense of protection. Crawling through thickets of beech may be abhorrent to foresters, but it is a sanctuary to me. These tenacious, tough trees take care of the soil with blankets of the thickest mulch you will find in the wild. They offer a limitless supply of carbon, beauty, and food. I hope that you, too, can appreciate this unique and resilient ally, the beech.

# Afterword
## Leaves of the Same Tree

It has been said by sages of ancient times that humans are leaves of the same tree, or fingers of the same hand. We see each other as different beings, but we are all physically connected from the same source. This does not go far enough. It is not just humanity, but all of creation that is one. All of existence is constantly melting and mixing into itself. At the core we can break things down into elements like carbon, hydrogen, oxygen, and many others. Everything in the world will become stardust. The Universe does not favor one side of itself more than another. When we learn of spinning galaxies filled with billions of stars, how can we believe that one country is more important than another? Why do we even have countries at all? We are riding a living planet through space, and we are all on the same ride together.

We all come from the same place and we all are headed to the same place. All of us: trees, people, rocks, water, animals. Not so long ago in history, people thought slavery was okay. It took a while for the mainstream to accept that all people are equal. It will take a while longer for people to accept that all of life is equal. All of life wants to live and is made of the same elements. Consciousness blows through all of the beings in the Universe, whether we recognize it or not. It is a prejudice to say that our way of perception is superior to another's. It has been proven by scientists that plants have a sense of hearing and a memory. They do not have a brain like ours, but they clearly have some form of consciousness. Plants share this world with us. They strive to express themselves and they have life stories filled with joy and tragedies. These sentient beings are built to work with us. If we can just notice, we will see that our world is filled with powerful allies: The trees of power are just some of them.

Life is eating itself everywhere, devouring bodies of plants, animals, and entire worlds. Supernovas sing out in creation and still people will say that God doesn't exist. They will say, "But look at all the bad stuff." So what? The Universe is not scared of pain or awkwardness or horror or anything. The Universe lives inside you and in everything you see. Cells, atoms, electrons, mitochondria are all dancing, making noise, being.

You can draw lines on a map, but the Universe will erase them in time. Your country is not any more important than other places. We are all leaves of the same tree. Beings lost in a bewildered, living dream. We are the play of existence. The strange workings of a Universe that enjoys experience.

Infinity is in all directions from where you sit, even down through your center. You are awake in a dream of incomprehensible immensity.

I watch my two-year-old son build block towers only to knock them down a minute later. He enjoys the action. It is the same with our work here on Earth. It will all be knocked down. Every species will go extinct, every tree will die, every person will vanish, melting into the oblivion of existence. We are drops of water disappearing into the ocean. Our work is for beauty and experience. When we forget the totality of existence, we take ourselves seriously and believe strange things are important.

I know some people will read this and think, *What is the point?* There does not have to be a point. Have you ever heard a song that touched you so deeply, you felt a longing for something that you could not describe? Have you ever danced without thought, or laughed so hard your stomach hurt? There was no point to any of those things, but you would certainly do them again. It is the same with working with trees. You will feel joy, frustration, inspiration, and wonder. Existence is beyond the scope of the human mind. It is for you to wonder about but never understand. If you could understand it, then it would not be so great. It would have to be made smaller to fit into your mind. Let it be great. Honor it, respect it, love it, laugh at it, spit on it, and roll over asleep with it. Life is for living.

There are many traps for humans to fall into. Thought patterns that drive home feelings of despair and depression. See these traps for what they are—places that you don't have to go. Your life force is precious. You are alive inside an infinite Universe. While you have awareness, embrace it. Don't get bogged down with words of gloom or normalcy. There is nothing normal here. There are no dates or places. You are at the center of infinity.

# Afterword

The present moment always offers something new. Every breath can bring a new thought. This dream is too beautiful to waste being miserable.

The trees are here with us. They are our allies, but they are also fingers of the same hand. All of us creatures who live upon the surface of the Earth draw our power from the same connected elements. Water that passes through my body is the same water that is drunk and expelled by trees of ancient and future times. Carbon that builds my body is the same carbon that has always been on the Earth. Our cells are filled with the same mysterious energy that all of creation is. Offer gratitude and you will receive it. Offer complaints, and you will receive them. We participate in creation with every thought and action, every dream and word. There are choices available every second. Victims, conquerors, saints, and humble tree planters—you can do anything. The Universe is not limited in possibility and so neither are you. *Reality* is a relative term. It means nothing other than an agreement between people.

Know that when air passes through your body, it is ancient and everlasting. That breath will exist forever as it passes from one being to the next and eventually out into space, where it will be for billions of trillions of years. It will transform into infinite forms, and a part of you will always be in its essence. The totality of the Universe is one. Manifested forms will rise and fall forever like my son's block towers. It is the churning over and over of itself. Find peace in your breath and know that no matter what you do, say, or think, you cannot be separated. Your breath, water, carbon, and cells are leaves of the same tree, fingers of the same hand that holds existence together. Ride this wave, for you are one drop of water in an ocean of infinite immensity. You are held so safely in this dream of infinity that even if you go through hell and your body is ripped to a million shreds, you will still be part of infinity no matter what you do. You are a drop of water in the ocean. A voice among a chorus of a billion trillion singers. Listen to the song being sung by the world. We can only hear a little bit of this song, but it gives us a glimpse of the beauty and madness that infinity holds. Thank you for existence.

## A Vision for Working with Trees

Civilization is our creation. We can make it anything. Right now we use a tremendous amount of our resources to kill people and numb our minds. I can imagine at least some of this energy being used for working with trees. Humans are so ambitious and creative. With a shift of awareness, the future has a way to crack and shell shagbark hickories efficiently; persimmons stand guard over doorways; elderberries line creek banks that are terraced with living willow walls; biomass crops fill the highway median strips; schoolyards are home to orchards, gardens, and berry patches big enough to feed all the kids that attend; parks are planted with pecans and chestnuts; barren fields are transformed into hazel thickets; mulberries are available in every backyard. Our soils are held secure by the roots of magnetic trees, while our bodies are cooled by their shade, our souls nourished by their generosity, our lives filled with the work of growing, harvesting, and preparing.

Trees offer partnership freely: no strings attached, no bill in Congress, no debate. They are here with branches extended in generosity. If we just notice and accept their gifts, what a different world we would have. But it is not just a dream. This reality is attainable and is happening. Billions of people care for the Earth and humanity. Many folks are already busy planting, grazing, processing, cooking, growing, learning endlessly. There is an entire subculture obsessed with working with nature. These people are ravenous for information. They listen constantly to the story the plants tell us about soil and biodiversity. They work hard and start businesses, write books, and produce volumes of healthy food. In some neighborhoods it will seem you will be the only one who cares about trees. However, if you start planting or harvesting seriously, people will notice. You will find that you are not the only one. This movement builds exponentially. It is built upon inspiring action.

If you tell people to do something, they will not. If you shine your own light, people will be drawn in to its warmth. Happy Planting.

# ACKNOWLEDGMENTS

It seems impossible to express how much love and gratitude I feel for the people and entities in my life who have helped me write this book.

I give thanks to the creator of the Universe for the unfolding of endless dreams. What an unbelievable, mysterious work you have created. I am enraptured with you and all the manifestations of infinity. I am so grateful to be in a conversation with you every second of every day. I love you to no end.

I give thanks to the plant people for all of their forms that have lit a fire in my mind and heart. Your shapes, colors, and feelings have touched me deeply. I am bewildered at the beauty in your expressions. You have stopped me in my tracks countless times. It is you, the plants, who have been the greatest inspiration in the creation of this book.

I give thanks to all of humanity. I am amazed at how much you all deal with. Being human is not always easy as we work with the powerful forces of thought, emotion, dream, and survival. There are certain humans who have touched me especially deeply. I cannot hope to mention you all here, as there have been many of you that have added so much beauty to my journey.

Thank you Megan for being the amazing partner you are, and sticking with me through the spiraling waves of our life together. Thank you Mom for loving me unconditionally and teaching me to never put up roadblocks. Thank you Rafi for showing me that life is an adventure and for living without fear. Thank you Ben for sharing this life with me and all of our amazing conversations that have expanded my awareness. Thank you to my kids for spreading so much love and joy into my heart and teaching me what it means to care for another.

Thank you Aunt Barb and Grandpa for being the cornerstones of the family and giving us all a place to return to again and again. And thank you to all the Silvers and Shumways, as you visit me in my dreams today as you did when I was a kid. Daniel, Hila, Yael, Zach, Noah, Helena, Jake, Rachel and Joe, Sam, Jillian, Uncle Mitch, Charlie, Jeff, Anne, Aunt Michelle, Aunt Toby, Michael, Eric, Megan, I love you all.

Thank you to the people at Buffalo Field Campaign. You woke me up to what our civilization is doing to the planet and changed my life forever. I appreciate you keeping up the fight out there. Free the wild!

Thank you Brian Caldwell for being a friend and a mentor and for planting trees. Thank you Sean Dembrosky for being a great friend and ally.

Thank you Sam Thayer for your work that has inspired me to no end, and thank you for now being a part of my work.

I am very grateful for the hardworking people at Chelsea Green for giving me this opportunity to share my knowledge and visions with the greater world. Thank you.

# EXERCISES FOR INCREASING AWARENESS

Awareness is the key to nature appreciation and participation. The senses of domesticated humans are extremely dulled. Tuning in to these senses again brings a surprising amount of wonder and awe. Deep appreciation of nature will lead to the best work. The following exercises have helped me over the years. This is just a small sampling. If you are interested in going further, there are many good books and schools. Wilderness Awareness School offers a course called Kamana that is completed at home through workbooks. It is a pretty amazing setup.

As your awareness increases you may notice that you are being watched everywhere you go. Birds and animals are experts at hiding. They live all around us and are constantly aware of their surroundings. As we come tromping through a place, they become very still and wait for us to leave. I feel a silent joy as I notice them, and even when I don't I am grateful for their presence. This is not limited to wilderness areas. I have seen mink, foxes, owls, coyotes, raccoons, and many other creatures in the most densely populated suburbs and cities. But nature awareness goes far beyond just noticing animals. There is a presence that flows through the world. In stillness, it can be seen. While we are busy doing, it is as if we are talking to the world. When we are able to just be, it is as if we are listening to the world. To really listen just once can change a person forever.

## Find a Hair

Literally find a hair. Pick any spot: grassy lawn, middle of the woods, old field, anywhere you feel comfortable lying on your belly and looking for a hair. I've tried this exercise many times and always find a hair eventually. I don't crawl around, just stay in one spot and comb through the grass or leaf litter. Sooner or later, you'll find a hair. It might be from a squirrel, a possum, a dog, a human, a deer, a fox, or a number of other critters. There are so many hairs blowing in the wind. Looking for the hair takes patience; don't rush it. The key is to enjoy the process. You will begin to notice the tiniest details that have been woven into the ground you walk on. There are countless miniature creatures, a mosaic of shape, color, and texture. As you focus in on the details of the ground, you can see just how elaborate it is. When you look up from the ground, you can get a different perspective on just how incredible and huge the entire world is.

## Sit Spot

This is the most powerful nature awareness exercise to me. Find a spot outside, not too far from home, where you can sit in private. You may have to crawl into a thick shrub, or maybe you will sit against the wall of your house, or maybe you will have a spot on the edge of a field or deep in the woods. Wherever it is does not matter nearly as much as what you do when you're there and how often you're there.

A routine of being in your spot at the same time every day strengthens this activity. You can sit for five minutes or an hour. Twenty to thirty minutes is a nice stretch. Make yourself comfortable; bring something soft and dry to sit on. Sit very still and pay attention. Pay rapt attention. Watch everything, listen with alertness, breathe. You are a receiver, an antenna—take in as much as you can without moving. Stretch your senses, listen, smell, feel, and see. Eyes, ears, and nose open wide. Feel the air upon your skin.

For roughly the first 20 minutes of a sit spot, the animals and birds are waiting. They have all seen you enter. After a little while they will go back to their activities and will pay little attention to you most of the time if you are very still. I've had birds land on me and chipmunks scurry over my legs. You cannot imagine the adrenaline rush of being arm's reach away from a deer that doesn't know you're there.

I believe doing a sit spot every day is one of the healthiest things for the mind and soul. You will also learn a ton about nature and see things that you will want to keep to yourself and tell everyone about at the same time.

There will be times when you are bored. That is when you refocus your senses. Every distraction, every thought is a reminder to get back to the present moment. Stay present, one breath at a time, only one. Over time you'll be able to sit for longer and longer if you continue to practice.

# Exercises for Increasing Awareness

## Owl Eyes

This exercise shifts consciousness and expands awareness, literally. Unfocus your eyes. See the world with wide vision. To start, you can hold your hands out to the sides and wiggle your fingers. See how far you can spread your arms and still see your wiggling fingers. Do this with your arms up and down as well as side to side. You should be able to see the sky and the ground at the same time as well as close to 180 degrees in front of you. Keep this wide view of the world for as long as you like. Practice going in and out of owl eyes, from the wide out of focus and back to the focused tunnel vision. I think you will find that it is a very pleasant way to view the world.

You will see all movement out of the corner of your eyes, as it will really jump out. It feels especially peaceful to open owl eyes among leaves moving in a light wind, but it's nice to do everywhere. You will not only see movement that you would have otherwise missed (the flick of a bird's tail, the scurry of a mouse) but also see the whole differently. I can't explain it, but it's worth checking out.

Owl eyes can also help you see in the dark and find things you have lost. If you drop something on the ground and can't find it, try relaxing your eyes and seeing things wide. The thing you are looking for will sometimes appear.

## Fox Walk

Fox walking involves walking with little or no sound. It requires surrender. If you are trying to walk from point A to point B, then fox walking is all but impossible. It works best if you can be totally present and not interested in where you are going.

To fox walk, do not look at the ground. Use your owl eyes to notice obstacles. Use your feet to feel them. Keep your head up. Keep your weight on your back foot as you pick up and move your forward foot. Reach out with your forward foot while you are totally balanced on the back. You should be able to stop with your front foot in the air at any point because your weight is on the back foot. As your front foot reaches the ground, touch lightly with the outside of your foot. Slowly roll to the inside. Once your front foot has completely felt and covered the ground, then slowly shift your weight to it. It may take anywhere from 10 to 60 seconds to complete a single step.

Unless you are a Taoist master, some strong feelings of impatience will come up. Focus on the present moment and surrender to being exactly where you are. Keep the owl eyes open as you slowly move through the land. A few minutes of fox walking with owl eyes will open up an astounding shift in consciousness.

Just be sure you're in a safe place from other people. You will look pretty weird to strangers. I once had someone call the police on me when I was fox walking through the woods in a park.*

### Get in the Thicket

Almost anywhere in the world, you can find a place that has not been set foot on in decades. These thickets are all around us and are largely unnoticed. A true thicket cannot be crawled into on hands and knees. You have to slide on your belly to enter. Once you get into the thicket, you will find yourself in a sanctuary, a place where no one can find you. Getting into thickets is exciting all by itself. They are wonderfully peaceful and exhilarating refuges that are worth visiting whenever you need a break or inspiration.

---

* Note: It's much easier to fox walk barefoot or in soft-soled shoes.

# RECOMMENDED READING

*Tree Crops* by J. Russell Smith. Writing back in 1929 during a time of severe soil degradation in the United States, Smith gave the country a loud warning and a detailed solution. His vision of tree crops has inspired the planting of millions of trees.

*Uncommon Fruits for Every Garden* by Lee Reich. I have read through this book from cover to cover at least a dozen times. Reich's descriptions of fruit trees and bushes that are largely unknown are detailed and inspirational. The book includes practical information on cultivation and propagation.

*You Can Farm* by Joel Salatin. This book is a powerhouse of inspiration. Before reading it, I thought that all farmers were poor and worked all the time. After reading, I realized that this was just not true. Salatin outlines his philosophies of being a creative entrepreneur and working in harmony with the land. It never would have occurred to me to become a full-time farmer without reading this book.

*The Forager's Harvest*, *Nature's Garden*, and *Incredible Wild Edibles* by Sam Thayer. These three books are probably the best books ever written about plants in the world. Sam's books are about foraging, but go far beyond that. They take a deep look at ecology, identification, harvest, and preparation, and contain ethnobotany and many great personal stories. If the world were ending and I had to run and could only take one book with me, it would be *Nature's Garden*.

*Sportsman's Guide to Game Animals* by Leonard Lee Rue. This book is not just for hunters; it is for all lovers of wildlife. Rue spent his entire life studying wild animals through intimate observation. All of his books are outstanding and riveting. If you want to learn about the habits and lives of wild animals, this is the best book to read.

*1491: New Revelations of the Americas Before Columbus* by Charles Mann. This book completely blew me away. It tells a history that was almost erased, of ancient magnificent civilizations that flourished throughout North and South America.

*Trees of the Eastern and Central United States and Canada* by William Harlow. This is an outstanding tree ID book. Harlow has several other books on the subject, including volumes on identification of seeds and winter twigs.

*Newcomb's Wildflower Guide* by Lawrence Newcomb. This is the best guide for identifying wildflowers anywhere. It is the only identification book I've ever used that is easy to use by following the keys.

*Farmers of Forty Centuries* by F. H. King. Written over 100 years ago. King travels throughout China, Korea, and Japan studying how they had achieved sustainable agriculture. It is remarkable to read about how large families grew all of their food and made an income from half-acre plots of land that lost no fertility over thousands of years.

*The Holistic Orchard* by Michael Phillips. Outstanding book covering fruit growing with the perspective of boosting plant and soil health.

*Growing Hybrid Hazelnuts* by Philip Rutter, Susan Wiegrefe, and Brandon Rutter-Daywater. A practical guide for growing hazelnuts.

# RESOURCES FOR PLANT MATERIAL

**Twisted Tree Farm:** This is my nursery, so of course I recommend it. Located in Spencer, New York. www.twisted-tree.net

**Edible Acres:** Wide collection of edible perennials grown with love instead of chemicals. Located in Trumansburg, New York. www.edibleacres.org

**Burnt Ridge Nursery:** Amazing collection of fruit and nut trees as well as berry bushes. Located in Washington State. www.burntridgenursery.com

**Oikos Tree Crops:** Unique genetics of many tree crops and edible perennials. Located in Michigan. www.oikostreecrops.com

**Perfect Circle Farm:** Quality nursery specializing in fruit and nut trees in Vermont. www.perfectcircle.farm

**Cummins Nursery:** Grows a huge variety of grafted fruit trees in upstate NY. www.cumminsnursery.com

**Fedco Trees:** Wide selection of fruit, nut, shade, ornamental, and shrubs in Maine. www.fedcoseeds.com

**Forest Agriculture Nursery (Forest Ag):** A breeding nursery specializing in precocious, cold-hardy, pest and disease resistant, edible woody crops. www.forestag.com

**Growing Fruit:** Online forum dedicated to sharing information and plant material. www.growingfruit.org

**NAFEX (North American Fruit Explorers):** Group of people who share information and plant material. www.nafex.org

**Northern Nut Growers Association:** Group of people who share information and plant material. www.nutgrowing.org

**Sheffields Seed Company:** Sells tree and shrub seeds. www.sheffields.com

**Summergreen:** Central Ontario's tree crops nursery and mushroom spawn source. www.summergreenfarm.com

**Hardy Fruit Tree Nursery:** Fruit and nut trees for extremely cold climates. Located in Quebec. www.hardyfruittrees.ca

# NOTES

### Chapter 1: Life Fountains
1. 4 Pour 1000, www.4p1000.org.

### Chapter 2: Planting
1. Charles Mann, *1491* (New York: Vintage Books, 2005).

### Chapter 7: Chestnut
1. Sandra Anagnostakis, *Saving the Ozark Chinquapins* (New Haven, CT: Connecticut Agricultural Experiment Station, n.d.), http://www.ct.gov/caes/lib/caes/pdio/documents/presentations/ozark_chinquapins__sla.pdf.
2. Noel Riggs, "FAOSTAT Data, 2005," Food and Agriculture Organization of the United Nations, http://www.fao.org/faostat.
3. Susan Freinkel, *American Chestnut: Life, Death, and Rebirth of a Perfect Tree* (Los Angeles: University of California Press, 2007).
4. Douglas Tallamy, *Bringing Nature Home* (Portland, OR: Timber Press, 2007).
5. J. Russell Smith, *Tree Crops* (New York: Harper and Row Publishers, 1929), 126.

### Chapter 9: Poplar
1. Peter Greatbatch, *The Practical Guide to Renewable Energy Using Hybridized Hardwoods* (Frisco, TX: Greatbatch, 1982).

### Chapter 10: Ash
1. "About Emerald Ash Borer," Emerald Ash Borer Information Network, www.emeraldashborer.info/about-eab.php.
2. Monitoring and Managing Ash, www.monitoringash.org.
3. Alexander Martin, Herbert Zimm, and Arnold Nelson, *American Wildlife and Plants* (New York: Dover Publications, 1951), 359.

### Chapter 11: Mulberry

1. Samuel Thayer, *Incredible Wild Edibles* (Bruce, WI: Forager's Harvest Press, 2017), 234.

### Chapter 12: Elderberry

1. Samuel Thayer, *Nature's Garden* (Birchwood, WI: Partners/West Book Distribution, 2010), 399.
2. Martin, Zimm, and Nelson, *American Wildlife and Plants*, 362.

### Chapter 14: Hazelnut

1. "Hazelnut Production," Food and Agriculture Organization of the United Nations, www.fao.org/docrep/003/x4484e/x4484e03.htm.
2. Peter Schwartzstein, "This Small Turkish Town Grows a Quarter of the World's Hazelnuts," *Quartz*, August 22, 2015, https://qz.com/483551/this-small-turkish-town-grows-a-quarter-of-the-worlds-hazelnuts.

### Chapter 16: Beech

1. Martin, Zimm, and Nelson, *American Wildlife and Plants*, 306.
2. Courtney LaMere, "Adirondack Black Bears," *New York State Conservationist*, April 2013, http://www.dec.ny.gov/docs/administration_pdf/0413adkblackbears.pdf.
3. Randall Morin, "Spread of Beech Bark Disease in the Eastern United States and Its Relationship to Regional Forest Composition," *Canadian Journal of Forest Research* 37 (2007): 726–36, https://www.nrs.fs.fed.us/pubs/jrnl/2007/nrs_2007_morin_001.pdf.
4. Jennifer Koch et al., "Screening for Resistance to Beech Bark Disease: Improvements and Results from Seedlings and Grafted Field Selections," in *Proceedings of the Fourth International Workshop on the Genetics of Host-Parasite Interactions in Forestry* (Albany, CA: USDA Forest Service, 2012), https://www.fs.fed.us/psw/publications/documents/psw_gtr240/psw_gtr240_196.pdf.

# INDEX

Note: Page numbers in *italics* refer to photographs.

abrasion, of seeds, 79
air, in soils, 54
air layering, 97–98
air-pruned beds
    for hickory, 83, 205
    principles of, 80–83, *81, 82*
    for seedling protection, 85
Akane apple, 143
Albers, Carl, 231
Allegheny chinquapin (*Castanea pumila*)
    for livestock, 122
    overview, 104, *104*
alley cropping, 125, 127
Amazonian civilizations, 12, 50–51, *52*
ambrosia beetles, 128–29
American beech (*Fagus grandifolia*), 236
American chestnut (*Castanea dentata*)
    overview, 103
    saga of, 108–14, *109–13*
American Chestnut Cooperators' Foundation, 113–14
American Chestnut Foundation, The (TACF), 112–13
American elderberry (*Sambucus canadensis*), 185
American hazel (*Corylus americana*), 210–11
Anagnostakis, Sandra, 105, 111, 130
ancient agriculture, landscape management, 11–12
annual agriculture, environmental concerns, 6, 8, *9*
Antonovka rootstock series, 145–46
apple (*Malus* spp.), 135–156
    air-pruned beds for, 83
    commercial possibilities, 154–56, *155*

domesticated, 138–39
grafting, 146–47
overview, 135–36
pests and diseases, 139, 152–54, *152*
rootstocks, 144–46
seedling apples, 147–49, *149*
species profiles, 140–41, *140*
stool layering, 99
uses of, 150–52, *151*
varieties, 141–44, *142*
wild, 136–38, *137*
wildlife value, 138, 149–150
wood of, 150
apple borer, 152–53, *152*
*Apple Grower, The* (Phillips), 156
applesauce, 151–52, 154–55
apple scab, 153
aquatic plants, as mulch, 46
Arbor Day Foundation, 213
ash (*Fraxinus* spp.), 165–172
    commercial possibilities, 171–72
    pests and diseases, 167–69, *168*
    species profiles, 166–67, *167*
    wildlife value, 170–71
    wood of, 169–170, *170, 171*
ash blight, 169
Ashmead's Kernel apple, 143
Ashworth, Fred, 212
Asian chestnut gall wasp, 129
aspen, bigtooth (*Populus grandidentata*), 162, *164*
aspen, quaking (*Populus tremuloides*), 41, 157–160, *158*
atmospheric carbon dioxide, tillage concerns, 6–7, *8*
awareness exercises, 251–54

bacterial vs. fungal soils, 41–42
Badgersett Research Corporation, 212
bareroot trees
    benefits of, 62–66, *63*, *66*
    heeling in, 35, *35*
    importance of keeping roots moist, 34–35
bare soil, concerns with, 6–7, *9*
bark
    as mulch, 45
    for tree identification, 26–27
bark grafting, 90, *90*, 94
baseball bats, ash for, 170
basket making, black ash for, 167
beaked hazel (*Corylus cornuta*), 211
beech (*Fagus* spp.), 235–244
    beech bark disease, 241–43, *242*
    commercial possibilities, 243
    growth habit and ecology, 237–38, *237*, 244
    nuts, 239–240, *240*
    propagation, 243
    species profiles, 236
    wildlife value, 240–41
    wood of, 238–39
beech bark disease, 241–43, *242*
beeswax, for grafting cuts, 92
beetles, ambrosia, 128–29
bench grafting
    apple, 146
    defined, 88
    mulberry, 92, 179
    scionwood for, 87
    storage of grafts, 92
    weather concerns, 92–93
    whip-and-tongue grafts for, 89
    *See also* grafting
berms and swales, 56–57, *57*
berries
    hardwood cuttings, 68, 70
    in the shade, 58
    tip layering, 97
    *See also* elderberry (*Sambucus* spp.); mulberry (*Morus* spp.)

Bhagwandin, Annie, 121
biennials, taprooted, for soil aeration, 61
big bug mite, 220–21
bigtooth aspen (*Populus grandidentata*), 162, *164*
biochar, 50–52
bird concerns
    chestnut seed propagation, 132
    hazelnut seed propagation, 221
    seed nuts, 83–84
Bird-X noise machines, 84
bitternut (*Carya cordiformis*), 196–98, *197*, *198*, 208
black ash (*Fraxinus nigra*), 167
black locust (*Robinia pseudoacacia*), 223–234
    chestnut vs., 116
    commercial possibilities, 232–33, *233*
    ecological effects of, 223, 224–25, 231–32, 233–34
    growth and flowering of, 224–26, *226*
    honey locust vs., 223–24
    nitrogen-fixing capability, 61
    pests and diseases, 227–28, *228*
    as pioneer species, 41, 223, 224, 231–32
    propagation, 229–230, *229*
    wildlife value, 228–29, 231, 232
    wood of, 226–27, 230–31, *230*, 232–33, *233*
black mulberry (*Morus nigra*), 176
Black Oxford apple, 143
blood, as fertilizer, 48–49
blue jays, protecting seed nuts from, 84
Bob Gordon elderberry, 192
boiled chestnuts, 119
bonemeal, as fertilizer, 49
book recommendations, 255–56
bottom heat, 71, 73, *73*, 75–76, *76*
box cutters, for grafting, 86
Bramley's Seedling apple, *142*, 143
breeding of trees

# Index

apple, 144–45
  benefits of, 27–29
  black locust, 230–31
  chestnut, 111, 112–14, *112, 113*
  elderberry, 192
  hazelnut, 210, 211–13, *214*
  hickory, 195, 199, 205–7, *206*
  hybrid swarm theory, 212
Budagovsky rootstock series, 145
budding, 90–91, *91*
Bull, Ephraim, 29
Burbank, Luther, 29
burdock, for soil aeration, 61
Burnham, Charles, 112
Burnt Ridge Nursery, 257
business opportunities. *See* commercial opportunities
Bussey, Daniel, 142

Cahokia (ancient city), 11
Caldwell, Brian, 113
callusing
  of cuttings, 68, 74, 75, *76*
  of grafts, 86
  weather concerns, 92
Calville Blanc d'Hiver apple, 143–44
candied nuts
  chestnuts, 122
  hickory, 204, 207
carbon, soil. *See* soil carbon
carbon cycle, 9–10
cardboard and paper mulch, 46
caretaking of nature, 10–14, *13*
*Carya* spp. *See* hickory (*Carya* spp.)
*Castanea* spp. *See* chestnut (*Castanea* spp.)
*Castanopsis* spp., 108
cedar apple rust, 153–54
chestnut (*Castanea* spp.), 103–34
  American chestnut saga, 108–14, *109–13*
  commercial possibilities, 133–34
  corn vs., 123–25
  cultivation, 125–27, *125*
  hybridizing, 29–30

  nuts, 109–10, 116–123, *116, 119, 120*
  pests and diseases, 127–130
  in pit-and-mound landscapes, 55–56, *56*, 126
  propagation, 83, 131–33, *132*
  species profiles, 103–8, *104, 105, 107*
  wildlife value, 109, 114–15, 123, 125
  wood of, 109, 115–16
  *See also specific types*
chestnut blight (*Cryphonectria parasitica*)
  attempts to overcome, 111–12, 129–130
  discovery and spread of, 110–11
  introduction to Europe, 106
  Japanese chestnut resistance to, 30
*Chestnut Cookbook, The* (Bhagwandin), 121
chestnut knives, 118–19
chestnut weevils, 127–28
China
  chestnut production, 106, 107
  use of silt traps, 52
Chinese chestnut (*Castanea mollissima*), 106–7
chip budding, 90
cider making, 151, 154, *155*
cleft grafting, 89–90, 94
climate change
  soil carbon benefits, 8
  tillage concerns, 6–7, 8
cloning, defined, 67
  *See also* cuttings
clover, nitrogen-fixing capability, 5, 60
cold periods, for sprouting. *See* stratification
colonialism, impact on food supply and landscape, 11–12
comfrey
  as fertilizer, 49–50
  as mulch, 45–46
commercial opportunities
  apple, 154–56, *155*
  ash, 171–72

commercial opportunities (*continued*)
  beech, 243
  black locust, 232–33, *233*
  caretaking aspects of, 13–14
  chestnut, 133–34
  earning money from trees, 31–32
  elderberry, 188, 193
  hazelnut, 212, 213–14, 219
  hickory, 207–8
  mulberry, 184
  poplar, 163–64, *164*
compost
  for layering, 98, *99*
  as mulch, 44
compost tea, as fertilizer, 49
Connecticut Agricultural Experiment Station, 105, 111, *112*, 129–130
controlled crosses, 29
coppicing
  apple, 150
  ash, *170*, 171
  beech, *237*, 238–39, 240, 243
  chestnut, 113, 115
  eastern cottonwood, 161
  hazelnut, 215
  mulberry, 176
  poplar, 161, 164
corn, chestnuts vs., 123–25
corn grinders, for chestnuts, *120*, 121
*Corylus* spp. *See* hazelnut (*Corylus* spp.)
cottonwood, eastern (*Populus deltoides*), 160–61, *161*
crab apple
  species profiles, 140–41
  wildlife value, 149
cracking/shelling nuts
  beechnuts, 239
  chestnuts, 119–120
  hazelnuts, 217–18
  hickory nuts, 202, *202*, *203*, 207
crosses, controlled, 29
*Cryptococcus fagisuga*, 242
Cummins, Jim, 144–45
Cummins Nursery, 257

curing, chestnuts, 117–18
cuttings, 67–76
  beech, 243
  bottom heat, 71, 73, *73*, 75–76, *76*
  cuttings, *69*, *70*, 71
  elderberry, 189
  hardwood cuttings principles, 68–70, *69*, *70*
  mist system, 72–74, *73*
  mulberry, 178–79
  overview, 67–68
  root cuttings principles, 70–71, *71*
  rooting hormone, 71, 74–75
  softwood cuttings principles, 71–72
  soil/growing medium, 76
  *See also* hardwood cuttings; root cuttings; softwood cuttings

Davebilt nutcrackers, 119–120, 217
decanting, for cleaning seed, 180
deer
  chestnuts for, 123
  elderberry protection from, 191
  protecting newly planted trees from, 36–37
dehesa system, 124–25
Dembrosky, Sean, 59
depth for planting
  sprouted seeds, 80
  trees, 36
direct seeding
  chestnut, 133
  hickory, 205
  overview, 83
dock spp., for soil aeration, 61
Dolgo crab apple, 142
dormancy, for planting trees, 62, 64, 65–66
drawing practice, for tree identification, 25–26
dried blood, as fertilizer, 48–49
drying
  apples, 152, 154–55
  chestnuts, 119–121, *120*

# Index

hickory nuts, 202
dwarf Giraldi mulberry, 183
dwarfing rootstocks, apple, 144

earthworks projects, 52–57, *55*, *56*, *57*
eastern cottonwood (*Populus deltoides*), 160–61, *161*
eastern filbert blight, 220
economic opportunities. *See* commercial opportunities
ecosystems, managed
    enhancing biodiversity, 18
    overview, 13–14, *13*
    soil, 38–39
    working through inspiration, 17–18
Edible Acres Nursery, 59–60, *59*, 257
elderberry (*Sambucus* spp.), 185–193
    commercial possibilities, 188, 193
    flowers of, 188, *188*
    fruit of, 188
    harvesting, 189
    planting, *190*
    in polycultures, 59, *59*
    propagation, *69*, *70*, 189–190
    species profiles, 185–86
    stem productivity, 186–87, *186*
    varieties, 191–92, *191*, *192*
    wildlife feed and habitat, 193
    wood of, 187
emerald ash borer, 168–69, *168*
energy for work, 20–21
environmental movement
    energy for work, 20–21
    need for positive action, 14–15
erosion
    cottonwoods for control of, 163
    tillage concerns, 6
European beech (*Fagus sylvatica*), 236
European chestnut (*Castanea sativa*), 105–6
European colonialism, impact on food supply and landscape, 11–12
European elderberry (*Sambucus nigra*), 185–86

European hazel (*Corylus avellana*), 210
Everloving Mulberry mother tree, 24–25, 183
Eve's Cidery, 139, *142*, *151*, *155*, 233
exercises for increasing awareness, 251–54

*Fagus* spp. *See* beech (*Fagus* spp.)
*Farmers of Forty Centuries* (King), 52, 256
Fedco Trees, 257
fencing, deer, 36–37
fertilizers
    burned hazel shells as, 219
    high-nitrogen, 41
    at planting, 47–52
field grafting, defined, 88
    *See also* grafting
field guides, for identification of trees, 26
filing, of seeds, 79
find a hair exercise, 251–52
fireblight, 153
firewood
    ash, 166, 170, 171
    beech, 238–39
    black locust, 224, 227, 232, 233
    hazelnut, 219
    hybrid poplars, 162
    mulberry, 177
fish emulsion, as fertilizer, 49
flours/meals
    chestnut, 120–22, *120*
    hazelnut, *218*, 219
flowers
    black locust, 225–26, *226*
    elderberry, 188, *188*
    hazelnut, 215–16, *216*
food supply, industrial agriculture concerns, 11
*Forager's Harvest, The* (Thayer), 255
ForestAg nursery, 257
*1491: New Revelations of the Americas Before Columbus* (Mann), 12, 255
fox walk exercise, 253–55

*Fraxinus* spp. *See* ash (*Fraxinus* spp.)
frost/freezing concerns
    apple crops, 137
    black locust seedlings, 229–230
    grafts, 92
    hazelnut crops, 214, 216
    mulberry seedlings, 181
    seeds, 78, 80
fruit processing
    apple, 154–55
    mulberry, 182
fuel value. *See* firewood
fungal vs. bacterial soils, 41–42

Gagliano, Monica, 21
garland crab apple (*Malus coronaria*), 140
genetic engineering, hybrid poplars, 163
Geneva rootstock series, 144–45
get in the thicket exercise, 254
gibberellic acid, 79
glyphosate (Roundup), 46–47
*Golden Guide to Trees of North America*
    (Golden Field Guides series), 26
Golden Russet apple, 143
grafting, 86–94
    aftercare, 93
    apple, 146–47
    beech, 243
    chestnut, 131
    materials for, 86–88, *87*
    mulberry, 179
    steps in, 88–93, *89, 90, 91*
    top-working older trees, 93–94, *93*
grafting knives, 86–87
grafting tar, 92
grafting tools, 86
graft unions
    burying, 36
    formation of, 86
grain equivalents in trees, 6
grass clippings, for mulch, 42, 43
gratitude for nature's bounty, 18–19
Graves, Arthur, 111
Greatbatch, Peter, 163

green ash (*Fraxinus pennsylvanica*), 166, *167*
grinding
    chestnuts, 120–21, *120*
    hazelnut press cakes, 219
Growing Fruit online forum, 257
*Growing Hybrid Hazelnuts* (Rutter et al.), 256
growing mediums
    for air layering, 97–98
    for cuttings, 76
    for mound layering, 96–97, *96*
    for seeds, 78
    for stool layering, 98–99, *98*
guano, as fertilizer, 48

hard cider, 151, 154, *155*
hardening off
    mist system considerations, 74
    venting tree tubes for, 37
    water concerns, 34
hardwood cuttings
    elderberry, 189
    mulberry, 178, 179
    principles of, 67, 68–70, *69, 70*
Hardy Fruit Tree Nursery, 257
Harlow, William, 26, 255
harvesting
    chestnut, 117
    elderberry, 189
    hazelnut, 216–17, *217*
    hickory, 200–201
    joy of, 3–4
    layers, 99
    mulberry, 182
Havahart traps, 84–85
hay, as mulch, 45
hazelnut (*Corylus* spp.), 209–22
    air-pruned beds for, 83
    commercial possibilities, 212,
        213–14, 219
    cultivation, 214–16, *215, 216*
    harvesting and processing, 216–19,
        *217, 218*
    major growing regions, 213–14
    overview, 209–10, *209*

# Index

pests and diseases, 220–21, 222
propagation, 221–22, 222
shells and wood value, 219
species profiles, 210–13, 212, 213
wildlife value, 220
hazelnut weevil, 221
heart rot (*Phellinus robiniae*), 228
heat mats and cables, 71, 73, 73, 75–76
heeling in, 35, 35
Hemlock Grove Farm, 9, 113, 116, 130
Henry chestnut (*Castanea henryi*), 107–8
herbicides, dangers of, 46–47
hicans, 206
hickory (*Carya* spp.), 194–208
    air-pruned beds for, 83
    commercial possibilities, 207–8
    harvesting, 200–201
    processing of nuts, 201–4, 202, 203, 207
    propagation, 204–5
    selection and breeding considerations, 205–7, 206
    species profiles, 194–99, 195, 197, 198
    wildlife value, 199–200
    wood of, 200
hickory brew/milk, 202–4, 207
*Holistic Orchard, The* (Phillips), 156, 256
honey locust (*Gleditsia* spp.), 223–24
horizontal planting, 98, 98
hot-water treatment
    black locust seed sprouting, 229
    chestnut weevil control, 128
Hudson's Golden Gem apple, 143
human-nature relationships
    caretaker role, 10–14, 13
    energy for work, 20–21
    gratitude for nature's bounty, 18–19
    joy of, 4–5
    love for nature, 10, 12
Hungary, black locust breeding, 231
husking/hulling
    beechnuts, 239
    hazelnuts, 211, 217
    hickory nuts, 198, 201–2

hybrid hazels, 211–13, 212, 213, 215, 221
hybridizing trees, 29–31
hybrid poplars, 162–63
hybrid swarm theory, 212

identification of trees
    mother trees, 23–25, 24
    skills for, 25–27
    techniques for, 27
Illinois Everbearing mulberry, 183
*Illustrated History of Apples in the United States and Canada, The* (Bussey), 142
income-earning opportunities. *See* commercial opportunities
*Incredible Wild Edibles* (Thayer), 176, 255
industrial agriculture, impact on food supply, 11
ink disease (*Phytophthera cinammomi*), 130
inoculants
    for encouraging fungal activity, 42
    for nitrogen-fixing, 60
interconnectedness of life, 245–47
interplanting
    chestnuts, 127
    principles of, 58–62, 59
invasive species concerns
    beech, 238
    black locust, 231–32, 233–34
    mulberry, 174
Italy, hazelnut production, 214

Japanese chestnut (*Castanea crenata*), 106, 107
Jensen, Derrick, 17
Jolicoeur, Claude, 151
Jonagold apple, 142
*Juniperus* spp., 153–54

Keepsake apple, 143
King, F. H., 52, 256
Kingston Black apple, 143
knives
    chestnut, 118–19
    grafting, 86–87

267

Kokuso mulberry, 183

LaMere, Courtney, 241
layering
    apple rootstocks, 146
    hazelnut, 222
    mulberry, 179–180
    principles of, 95–99, *95*, *96*, *99*
leafy plants, as mulch, 45–46
leaves
    as indicator of soil nutrient deficiencies, 40
    for tree identification, 25–26
Liberty apple, 142
lingering ash, 168
little leaf mulberry (*Morus microphylla*), 176
livestock feed
    black locust, 229
    chestnuts, 122–23, 124–25
    hazelnut press cake, 219
    mulberry, 184
locust borer (*Megacyllene robiniae*), 227–28, *228*
locust leaf miner (*Odontota dorsalis*), 228

Malling rootstock series, 145
*Malus sieversii*, 141
*Malus* spp. *See* apple (*Malus* spp.)
MaMA (Monitoring and Managing Ash), 168–69
Mann, Charles, 12, 255
manure
    in compost tea, 49
    as fertilizer, 48
    as mulch, 43–44
    for pH balancing, 214
Marge elderberry, 192
Master Nut Cracker, 202, *202*
meal, chestnut, *120*, 121–22
medium/soil, for cuttings, 76
microorganisms, in the soil, 39
micro topography of forests, 53
Millican, Marge, 192

mint, pollinator benefits, 61–62
mist systems, 72–74, *73*
mold concerns
    chestnut seed propagation, 131
    hardwood cuttings, 69
    seed storage, 78
Monitoring and Managing Ash (MaMA), 168–69
*Morus* spp. *See* mulberry (*Morus* spp.)
mother trees
    collecting seeds from, 28
    identification of, 23–25, *24*
mound layering, 96–97, *96*
mowing
    of apple hedgerows, 148
    for deer travel corridors, 37
    grass clippings from, 43
    for protection against voles, 38
    for weed control, 42
mulberry (*Morus* spp.), 173–184
    commercial possibilities, 184
    Everloving Mulberry mother tree, 24–25, 183
    harvesting and processing, 182, *182*
    overview, 173–75, *173*
    pests and diseases, 180
    propagation, 177–181, *180*, *181*
    shade tolerance, 60
    species profiles, 175–76
    varieties, 182–84
    wood of, 176–77
mulch
    benefits when planting, 34
    for encouraging fungal activity, 41, 42
    interface with soil, 41
    in nursery beds, *40*
    for protection of fertilizer, 47–48
    for weed control, 42–46, *42*
mushroom production
    beech logs for, 243
    chestnut logs for, 116
mycorrhizal fungi, 41
*Mycorrhizal Planet* (Phillips), 156

# Index

NAFEX (North American Fruit Explorers), 257
nature-human relationships. *See* human-nature relationships
*Nature's Garden* (Thayer), 189, 255
netting, for seed nut protection, 84
*New Cider Maker's Handbook, The* (Jolicoeur), 151
Newcomb, Lawrence, 256
*Newcomb's Wildflower Guide* (Newcomb), 256
New York State Agricultural Experiment Station (Geneva, New York), 141
New York Tree Crops Alliance, 128, 133, 218
Nitka hazelnut, *212*, 213, *213*
nitrogen, as nutrient, 40–41
nitrogen-fixing species, *5*, 60–61, 224
North American Fruit Explorers (NAFEX), 257
Northern Nut Growers, 257
Northern Spy apple, 142
nursery beds, *40*
nursery stock
    apple, 154
    black locust, 233
nutcrackers, Davebilt, 119–120, 217
nut oils
    bitternut, 197, *198*
    hazelnut, 218–19, *218*
    hickory, 204, 208
nuts
    beech, 239–240, *240*
    chestnut, 109–10, 116–122, *116*, *119*, *120*
    hazelnut, 216–19, *217*, *218*
    hickory, 201–4, *202*, *203*
nut wizards, 117, 201

Oikos Tree Crops, 257
oilnut (bitternut) (*Carya cordiformis*), 196–98, *197*, *198*, 208
oils, nut. *See* nut oils
Ordu, Turkey, hazelnut production, 213–14
organic matter, in soil
    benefits of, 9–10, 34
    loss of, 7–8
organic production
    apple, 138–39
    black locust, 232–33, *233*
    chestnut, 127, 128, 133
Oriental beech (*Fagus orientalis*), 236
ornamental beeches, 236
owl eyes exercise, 253
Ozark chinquapin (*Castanea ozarkensis*)
    Asian chestnut gall wasp resistance, 129
    overview, 104–5, *105*

Pakistani mulberry, 183
paper and cardboard mulch, 46
parasitic fungi, 41
passenger pigeons, and beech, 241
pawpaw, air-pruned beds for, 83
pear, wild, 4
pecan (*Carya illinoinensis*), 199
peeled chestnuts, 120–21
pellicles, chestnut, 106, 120
perennials, shade, 62
Perfect Circle Farm, 257
perry (pear wine), 4
persimmon, air-pruned beds for, 83
pests and diseases
    apple, 139, 152–54, *152*
    ash, 167–69, *168*
    beech, 241–43, *242*
    black locust, 227–28, *227*
    chestnuts, 127–130
    hazelnut, 220–21, *222*
    mulberry, 180
    protection for newly planted trees, 36–38
Phillips, Michael, 156, 256
pH of soil
    for chestnut, 126
    for hazelnut, 214, 221

pH of soil (*continued*)
   overview, 39–40
   wood ashes effect on, 50
pignut (*Carya glabra*), 198–99
P-I hazelnut, 213
pine, air-pruned beds for, 83
pioneer species
   black locust, 41, 223, 224, 231–32
   overview, 41
   quaking aspen, 159–160
   white ash, 166
pit-and-mound landscapes
   benefits of, 53, 54, 55–57, *55*
   chestnut, 55–56, *56*, 126
planting, 33–66
   bareroot benefits, 62–66, *63*, *66*
   elderberry, *190*
   fertilizing, 47–52
   heeling in, 35, *35*
   interplanting, 58–62, *59*
   placing in hole, 36
   protection from pests, 36–38
   soil, 38–42, *40*
   uneven ground considerations, 52–57, *55*, *56*, *57*
   water, 34–35
   weed control, 42–47, *42*
planting depth. *See* depth for planting
plants as sentient beings, 21–23
pollination
   beech, 239
   chestnuts, 125–26, *125*
   controlled crosses, 29
   elderberry, 188, 191
   hazelnut, 215–16
   plants for encouraging pollinators, 61–62
polycultures. *See* interplanting
poplar (*Populus* spp.), 157–164
   commercial possibilities, 163–64, *164*
   hybrids, 162–63
   as pioneer species, 41
   propagation, 163

species profiles, 157–162, *158*, *161*
potted plants
   bareroot trees vs., 62–65
   chestnut, 132
   hickory, 205
   taprooted species, 80–81
*Practical Guide to Renewable Energy Using Hybridized Hardwoods, The* (Greatbatch), 163
pre-colonial abundance, 11–12
press cake, hazelnut, *218*, 219
Pristine apple, 143
profit-making opportunities. *See* commercial opportunities
propagation
   beech, 243
   black locust, 229–230, *229*, 233
   chestnut, 131–33, *132*
   elderberry, 189–190
   hazelnut, 221–22, *221*
   hickory, 204–5
   mulberry, 177–181, *180*, *181*
   poplar, 163
   *See also* cuttings; grafting; layering; seed propagation
pruning
   elderberry, 186–87, *186*
   hazelnut, 215
   roots, when planting, 36
   *See also* air-pruned beds
pyrethrum, 128–29

quaking aspen (*Populus tremuloides*), 41, 157–160, *158*

rake handles, white ash for, 169
ramial wood chips, 44
raw manure, as mulch, 43–44
reading recommendations, 255–56
red cedar (*Juniperus* spp.), 153–54
red mulberry (*Morus rubra*), 176
regenerative power of trees, 5–6
Reich, Lee, 255
resources for plant material, 257

# Index

roasting
    chestnuts, 118–19, *119*
    hazelnuts, 218
*Robinia pseudoacacia*. *See* black locust (*Robinia pseudoacacia*)
rodent concerns
    air-pruned beds, 82
    chestnut seed propagation, 131, 132
    seed nuts, 83–85
    seed storage, 78
root cuttings
    beech, 243
    black locust, 230
    elderberry, 190
    principles of, 70–71, *71*
root hairs, 38–39
rooting hormone, 74–75
roots
    bareroot tree benefits, *63*, 64
    placing trees into holes, 36
rootstocks
    apple, 144–46
    for grafting, 88
rot resistance
    black locust, 226–27
    chestnut, 109, 116
    mulberry, 177
round-headed apple borer, 152–53, *152*
Roundup (glyphosate), 46–47
Roxbury Russet apple, 143
Rue, Leonard Lee, 255
*Rumex* spp., for soil aeration, 61
Rutter, Philip, 112, 127, 212, 256
Rutter-Daywater, Brandon, 256

safety considerations, grafting, 87
Salatin, Joel, 255
*Sambucus caerulea*, 186
*Sambucus racemosa*, 186
*Sambucus* spp. *See* elderberry (*Sambucus* spp.)
sand
    for chestnut storage, 118
    for cutting propagation, 76
    for graft storage, 92
    for seed stratification, 78
saprophytic fungi, 41
sargent crab apple (*Malus sargentii*), 140, *140*
sawdust
    for chestnut storage, 118
    for graft storage, 92
    as mulch, 44–45, 180
    in stool beds, 98, *98*
    for storing hardwood cuttings, 69
scale insects, 242
scarification, 79
scionwood, 87–88, *87*
Scotia elderberry, 191
seasonal pools, in forests, 54–55
seed collection, for breeding, 28
seedling apples, 147–49, *149*
seed nuts, protecting from predators, 83–85
seed propagation, 77–85
    air-pruned beds, 80–83, *81*, *82*, 85
    apple, 148–49
    beech, 243
    black locust, 229–230, *229*, 233
    chestnut, 131–33, *132*
    hazelnut, 221–22, *221*
    hickory, 204–5
    mulberry, 178, 180–81, *180*, *181*
    protecting seed nuts from predators, 83–85
    scarification, 79
    stratification, 77–78, *77*, *79*
Seguin chestnut (*Castanea seguinii*), 107
sentience of plants, 21–23
separators, nut, 217–18
Serik, Steve, 116
serpentine layering, 97
sexual propagation, defined, 67
    *See also* seed propagation
shade perennials, 62
shade tolerance
    beech, 237–38, 239
    berry crops, 58, 60
    elderberry, 185, 191

shagbark hickory (*Carya ovata*), 194–95, *195*, *206*
Sheffields Seed Company, 257
shellbark hickory (*Carya lacinosa*), 195
shelling/cracking nuts
    chestnuts, 119–120
    hazelnuts, 217–18
    hickory nuts, 202, *202*, *203*, 207
shells, hazelnut, 219
shiitake production
    beech logs for, 243
    chestnut logs for, 116
shrubs, nitrogen-fixing, 61
Siberian crab apple (*Malus baccata*), 141
Sierra, Ricardo, 166
silt traps, 52
silvopasture, 124–25
simple layering, 96
sit spot exercise, 252
slug concerns, mulberry cultivation, 180
Smith, J. Russell, 6, 123, 255
snap traps, 85
soaking, of seeds, 79
softwood cuttings
    defined, 67
    elderberry, 189
    mulberry, 179
    principles of, 71–72
soil
    bacterial vs. fungal, 41–42
    bareroot tree benefits, 63
    for cuttings, 76
    preparation for planting, 38–42, *40*
soil carbon
    biochar, 50–52
    for microorganisms, 39
    tree crops and, 6–10, *9*
soy meal, as fertilizer, 49
spacing
    bareroot trees, 64–65
    chestnuts, 126–27
    seedling apples, 148

sphere of influence, working within, 15–17
*Sportsman's Guide to Game Animals* (Rue), 255
spring ephemerals, 62
stool layering
    apple rootstocks, 146
    mulberry, 179–180
    principles of, 98–99, *98*
storage
    apples, 150–51
    apple seeds, 148–49
    bench-grafted trees, 92
    chestnuts, 118
    hardwood cuttings, 69
    scions, 88
    seeds, 78
    trees grown in air-pruned beds, 82
Stoscheck, Autumn, *151*
stratification
    chestnut, 131
    hazelnut, 221, *222*
    hickory, 204
    principles of, 77–78, *77*, *79*
straw, as mulch, 45
sugar maple, 225
Summergreen Farm, 257
swales and berms, 56–57, *57*
Sweet 16 apple, 143
sweet chestnut. *See* European chestnut (*Castanea sativa*)
sweet-scented crab apple (*Malus coronaria*), 140
synthetic fertilizers, 48

TACF (The American Chestnut Foundation), 112–13
talking to plants, 22–23
taping grafting cuts, 91–92
taprooted species
    air-pruned beds for, 80–83, *81*, *82*, 85
    biennial, for soil aeration, 61
tar, for grafting cuts, 92
T-budding, 91, *91*

# Index

terra preta (fertile ground), 50–52
Texas mulberry (*Morus microphylla*), 176
Thayer, Samuel, 176, 189, 196, 211, 255
The American Chestnut Foundation (TACF), 112–13
thicket exercise, 254
tillage of soil, concerns with, 6–7, *9*
timing
    planting bareroot vs. potted trees, 65–66
    seed sprouting, 80
tip layering, 97
tongue cuts, 87, 89, *89*
topography of forests, 53
top-working older trees, 93–94, *93*, 146
transplanting
    air-pruned bed benefits, 80, 83
    apple seedlings, 149
    bareroot trees, 62, 64, 65
    black locust, 230
    elderberry, 189–190
    hardwood cuttings, 70
    hickory, 205
    softwood cuttings, 74
traps, for protecting seed nuts, 84–85
*Tree Crops* (Smith), 6, 123, 255
tree crops, soil carbon and, 6–10, *9*
    *See also specific types of trees*
*Trees of North America* (Golden Field Guides series), 26
*Trees of the Eastern and Central United States and Canada* (Harlow), 255
tree tubes, 37
tree wraps, 38
Turkey, hazelnut production, 213–14, 216
Turkish tree hazel (*Corylus colurna*), 211
twig/leaf boards, 26
Twisted Tree Farm, *40*, *132*, *137*, *149*, 257

*Uncommon Fruits for Every Garden* (Reich), 255
uneven ground, utilizing, 52–57, *55*, *56*, *57*

University of Missouri, 192, 193
U-pick mulberry business opportunities, 184
Upper Midwest Hazelnut Development Institute, 218
urine, as fertilizer, 48
US Department of Defense, 213
US Forest Service, 168

vernal pools, in forests, 54–55
vision for working with trees, 247–48
voles, protecting newly planted trees from, 37–38
Volusia soil, 53

walnut, air-pruned beds for, 83
Walsh, John, 120
water
    air-pruned beds, 82
    bareroot tree benefits, 64
    hardwood cuttings, 68
    planting trees, 34–35
    softwood cuttings, 72
wax, for grafting cuts, 92
weed control
    herbicide dangers, 46–47
    mowing, 42
    mulch, 42–46, *42*
weeping mulberry, 182–83
weevils, chestnut, 127–28
Weschke, Carl, 212
whip-and-tongue grafting, 87, 89, *89*, 94
white ash (*Fraxinus americana*), 166, 169–170, *170*, *171*
white mulberry (*Morus alba*), 175–76
white mulberry varieties, 183–84
white poplar (*Populus alba*), 162
Wickson crab apple, 143
Wiegrefe, Susan, 256
wild apples, 136–38, *137*
wilderness, misconceptions about, 11–12

wildlife feed and habitat
   apple, 138, 149–150
   ash, 170–71
   beech, 240–41
   bitternut, 198
   black locust, 228–29, 231, 232
   chestnut, 109, 114–15, 123, 125
   elderberry, 193
   hazelnut, 220
   hickory, 199–200
   mulberry, 174
   poplar, 159–160, 161, 162
wild pear, 4
Willamette Valley, hazelnut production, 214
wood
   apple, 150
   ash, 169–170, *170*, 171
   beech, 238–39
   black locust, 226–27, 230–31, *230*, 232–33, *233*
   chestnut, 109, 115–16
   elderberry, 187
   hazelnut, 219
   hickory, 200
   mulberry, 176–77
   poplar, 163–64, *164*
wood ashes, as fertilizer, 50
wood chip mulch, *40*, 44
wraps, for trees, 38
Wyldewood elderberry, 191–92

Yellowstone National Park, 16–17
York elderberry, 191, *191*
*You Can Farm* (Salatin), 255

Zarnowski, Jeff, 212, 213
Z's Nutty Ridge, 213, *215*

# ABOUT THE AUTHOR

**Akiva Silver** owns and operates Twisted Tree Farm, a homestead, nut orchard, and nursery located in Spencer, New York, where he grows around 20,000 trees a year using practices that go beyond organic. His background is in foraging, wilderness survival, and primitive skills. He has been observing nature intensively for the last 20 years, cultivating a deep appreciation for life. Akiva lives on his farm with his wife and three young children.

## About the Foreword Author

**Samuel Thayer** is an internationally recognized authority on edible wild plants who has authored two award-winning books on the topic, *Nature's Garden* and *The Forager's Harvest*. He has taught foraging and field identification for more than two decades. Besides lecturing and writing, Samuel is an advocate for sustainable food systems; he owns a diverse organic orchard and harvests wild rice, acorns, hickory nuts, maple syrup, and other wild products. He lives in rural northern Wisconsin with his wife and three children.

the politics and practice of sustainable living

# CHELSEA GREEN PUBLISHING

Chelsea Green Publishing sees books as tools for effecting cultural change and seeks to empower citizens to participate in reclaiming our global commons and become its impassioned stewards. If you enjoyed *Trees of Power*, please consider these other great books related to nature and ecology.

**GODS, WASPS AND STRANGLERS**
*The Secret History and Redemptive Future of Fig Trees*
MIKE SHANAHAN
9781603587976
Paperback • $14.95

**MESQUITE**
*An Arboreal Love Affair*
GARY PAUL NABHAN
9781603588300
Hardcover • $22.50

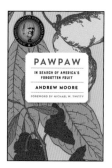

**PAWPAW**
*In Search of America's Forgotten Fruit*
ANDREW MOORE
9781603587037
Paperback • $19.95

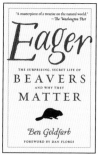

**EAGER**
*The Surprising, Secret Life of Beavers and Why They Matter*
BEN GOLDFARB
9781603589086
Paperback • $17.95

For more information or to request a catalog, visit **www.chelseagreen.com** or call toll-free **(800) 639-4099**.